THE BIG BOOK OF **CIDERMAKING**

THE BIG BOOK OF
CIDERMAKING

Expert techniques for **FERMENTING** *and*
FLAVORING *your favorite* **HARD CIDER**

CHRISTOPHER SHOCKEY **and** **KIRSTEN K. SHOCKEY**

PHOTOGRAPHY BY CARMEN TROESSER

Storey Publishing

*The mission of Storey Publishing is to serve our customers by
publishing practical information that encourages
personal independence in harmony with the environment.*

EDITED BY Carleen Madigan and Sarah Guare

ART DIRECTION AND BOOK DESIGN BY Carolyn Eckert

TEXT PRODUCTION BY Liseann Karandisecky

INDEXED BY Christine R. Lindemer, Boston Road Communications

COVER AND INTERIOR PHOTOGRAPHY BY © Carmen Troesser

ADDITIONAL INTERIOR PHOTOGRAPHY BY Mars Vilaubi, 45 b.,
138, 139, 201, 209 t., 238, 250, 255; © Andrii Chernov/Alamy
Stock Photo, 232; courtesy of Bill Bleasdale, 143; © Botanist &
Barrel, 190; © Francois de Melogue, 126; © Haritz Rodriguez/
Ciderzale.com, 166, 167; © Jon Arnold Images Ltd/Alamy
Stock Photo, 142; courtesy of Mt. Hood Organic Farms, 127
l.; courtesy of MUSE Marketing + Design, 231 l. & r.; courtesy
of Peter Mitchell, 291; courtesy of Trask Bedortha 230, 231
c.; © Yasmin Khajavi Photography, 127 r.

PHOTO STYLING BY Carmen Troesser

FOOD STYLING BY Christopher Shockey and Kirsten K. Shockey

ILLUSTRATIONS BY Alois Lunzer/Boston Public Library/Wikimedia
Commons, 86 and throughout; Carolyn Eckert, 19 and throughout; Grace Carter/Boston Public Library/Wikimedia Commons,
133; Ilona Sherratt, 30; Olive E. Whitney/Boston Public Library/
Wikimedia Commons, 48 and throughout

TEXT © 2020 BY Christopher Shockey and Kirsten K. Shockey
A portion of the preface was originally published in *Comestible*.

The information in this book is true and complete to the best of
our knowledge. All recommendations are made without guarantee
on the part of the author or Storey Publishing. The author and
publisher disclaim any liability in connection with the use of this
information.

Storey books are available at special discounts when purchased in
bulk for premiums and sales promotions as well as for fund-raising
or educational use. Special editions or book excerpts can also be
created to specification. For details, please call 800-827-8673, or
send an email to sales@storey.com.

Storey Publishing
210 MASS MoCA Way
North Adams, MA 01247
storey.com

Printed in China through World Print
10 9 8 7 6 5 4 3 2 1

Library of Congress Cataloging-in-Publication Data on file

Hard cider is an alcoholic beverage. Please consume alcohol and cannabis responsibly and be aware of any related laws in your area.

Read all instructions thoroughly before using any of the techniques or recipes in this book and follow all safety guidelines.

This publication is intended to provide educational information for the reader on the covered subject. It is not intended to take
the place of personalized medical counseling, diagnosis, and treatment from a trained health professional.

Dedicated to our kids and the beautiful people they share their lives with now. So much of this book is about the place where you grew up, and we hope that you will aways call it home.

What if you could help heal our environment in a small way, improve your health, and have fun through the drink?

CONTENTS

Foreword

My love affair with cider began way back in the late 1980s, when I first met Terry and Judith Maloney, who owned a little business in western Massachusetts called West County Winery. Today their son, Field, carries on the tradition at what has become West County Cider.

The Maloneys were among the first key players in what's come to be known as the cider renaissance in the United States. But in truth, in those early years of the revival, there were very few people making cider commercially — good, bad, or indifferent — and though there were plenty of indications that cider was growing in popularity, for years this growth was quite organic and slow, almost glacial in fact, and not always easy. Another pioneering cidermaker once told me that in the beginning he'd practically had to kneel on people's chests and pour his product down their throats. An exaggeration, to be sure, but the fact remains that Americans are still relearning the joys of cider, which in earlier times was this country's foundational beverage.

The thing that struck me about cider back in the 1980s and '90s was how convivial and genuine the people who made it were. The culture was different from that of high-end wines, and even friendlier than that of craft beer, which was booming at the time. There was nothing standoffish or proprietary about cider producers, at least not the ones who were growing and fermenting their own fruit and experimenting with old (and new) apple varieties to evaluate their relative worth. You could ask these people which apples they were using and see exactly how they made their cider, and no one was afraid you'd steal their trade secrets (as if they held the recipe for Coca-Cola or Kentucky Fried Chicken).

That's because these people understood that cider at its highest and best expression doesn't use a recipe, but a process. Also, they knew that all cider is local. By that I mean, even if I were to ferment the same juice from the same apples another cidermaker had used, I wouldn't end up with exactly the same cider. That's because, in addition to terroir (the well-known concept of the influence of such factors as climate, geography, and soils on fruit), a distinctive human element is also involved, one that contributes either a little or a lot to a particular cider. Vive la difference.

To prove this point, I once led a taste workshop for a group of 60 or 70 people where we tasted three single-variety ciders made by three different cideries from the famous Kingston Black apple. In fact, the apples used all came from the same orchard in western New Hampshire, so the terroir should have been the same. Yet each of these ciders was quite distinctive — all very well made and delightful, but all significantly different from one another.

Today I still find the same spirit of generosity and collegiality among cidermakers, both amateurs and pros. And both the quality and diversity of ciders are exponentially greater these days. Cidermakers are creating rosé cider made from red-fleshed apples, sour ciders, and all manner of specialty ciders that feature adjuncts or special ingredients — not just traditional barrel aging in oak or throwing some raisins into a New England–style cider.

Christopher and Kirsten Shockey have captured in this book all of the passion and joie de vivre that attracted me to cider so many years ago. In a friendly, straightforward manner, they describe the whole process of making cider on a variety of scales and present an impressive palette of options for neophytes, hobbyists, and even more experienced cidermakers. Their long experience with fermentation, in all its forms, provides a sound basis for success your first time out, and every time. And the information they give on making yeast cultures from foraged fruits and flowers puts an exciting and innovative local spin on this unique beverage — one that's had a proud yet humble pedigree and promises to have an even brighter future.

— **BEN WATSON**, author of *Cider, Hard and Sweet*

The Ark of Apples

This book, and our love affair with apples and cider, started in 1998 — the year we moved onto our smallholding and watched the dormant centenarian apple trees introduce themselves.

First came the pink swelling blossom buds, next the riot of white blossoms resplendent and humming with pollinators, then green leaves offering cool summer shade as the small fruits grew into the apples. We soon identified most of them — a Rome variety of some sort, something like a Granny Smith, a Golden Delicious, a Cox's Orange Pippin, a few towering Gravensteins, and one that was grafted to both Gravenstein and Red Delicious. We were overwhelmed by the quantity: boxes and baskets of apples were stacked along the wall in our small kitchen. Apples seemed to tumble every which way as we tried to make them into sauce, dried rings, steamed juice, pies, crisps, and dumplings. By the next year we had a cider press, and a few years later we were captivated by cider. Surrounded by vineyards, we thought we would be the first cider house in our area. As it turned out, sauerkraut got in the way, but that is another story.

Eager to learn as much as we could about growing apples for cider, we visited Nick Botner, described both as a hobby orchardist and a serious world-renowned botanical collector, at his farm in Yoncalla, Oregon, 2 hours north of our farm. We arrived, three of our four children in tow, one early November day, nearly 15 years ago. "Come into my farmhouse, we'll talk," Nick said as he invited the five of us in. His wife, Carla, sat us down to coffee and applesauce.

"What kind of apples do you recommend for hard cider?" Christopher ventured. We were sitting there gazing at him like initiates around a sage, waiting for the meaning of life. Or, at least the meaning of apples.

"There are a lot of great apples for cider," Nick said, and we both stared, pen in hand, waiting to scribble down the varieties that we'd never heard of, yet hoped to plant. He told us a good cider apple contributes to one or more of four components: color, flavor, body, or bouquet. He didn't drop any variety names though.

A good cider apple contributes to one or more of four components: color, flavor, body, or bouquet.

Nick started gathering apple varieties in 1976 simply by trading twigs with other apple growers from all over the world, like pen pals who shared pieces of scion wood instead of stories.

"Do you have the Redstreak?" Christopher asked hopefully. During the eighteenth century, this apple was believed to be the finest cider apple in England. At the time, cider made from the Redstreak commanded the highest prices. Its popularity had diminished by the end of the century and it's believed that viruses may have killed the remaining trees. Now the apple is rare, even thought to be extinct, as breeders are unsure if the claimed Redstreaks are indeed *the* Redstreaks.

"Yes, I believe I do," Nick said. "Would you like to see the orchard?"

The two of us nearly jumped out of our seats. We all put our rubber boots back on and traipsed through wet fall leaves down the knoll to his orchard. This was where Nick started gathering apple varieties in 1976 simply by trading twigs with other apple growers from all over the world, like pen pals who shared pieces of scion wood instead of stories.

Scion wood is a small dormant stick of a tree's new growth that can be grafted onto a rootstock of another apple tree to start a clone of the scion variety. This can happen over and over again. The buds carry the plant's knowledge — the genes needed for the desired variety, and when they "wake up" they grow upright, becoming the trunk of the new tree.

Nick and his worldwide network have protected and preserved apple varieties that would have otherwise died out. To put this into context, industrialized apple growing is like putting biodiversity through a funnel. There are about 7,500 named apple varieties in the world, yet in the United States only about 100 varieties are grown commercially. Apple seeds don't reproduce the same fruit as the apple that parented them; you never know what you will get. Each apple has between 5 and 12 seeds, and each of those seeds will grow into a different variety of tree. Think of all the apples from a single tree and you understand how an apple tree is a diversity-generating wonder. In the wild apple forests of Kazakhstan, no two varieties are the same, and many are only edible for birds. Through the centuries, humans have selected and cloned for flavor, grafting scion wood from the preferred variety onto rootstocks. Varietals are based on keeping a clone of the original tasty tree going — for centuries. And here lies the problem of the questionable Redstreak.

"Some of the ugliest ones taste the best," Nick told us.

SPICED RUSSET

Nick's orchard was then a world-class collection of humble trees holding over four thousand varieties of *Malus domestica* — the apple. Some varieties were so rare they were thought to be near extinction, and perhaps the only living clones of those varieties were in his orchard.

On that cool and gray November day, most of the trees were nearly bare of leaves, but they still held orbs of yellow, gold, green, russet, purple, almost white, and every hue of red imaginable. Some were huge — like garish Christmas ornaments hanging on limbs that bent under the weight — and there were tiny crabapples that hung in ruby red clusters like a loaded cherry tree. Carpets of interwoven wet leaves and fallen apples in all stages — from fresh to rotten — lay under trees, making compost to nourish the soil, the microbes, and the worms.

"Just pick anything and try it," Nick told us as he closed a gate behind us. "Some of the ugliest ones taste the best."

The first one Kirsten picked was a sensual crimson that drew us in like Snow White's poison apple. "What kind is this?" She asked as she bit into the crisp, cold flesh.

Nick carried a worn, thick three-ring binder, which held the key to the whole orchard; each row was numbered and cataloged on those loose-leaf typed pages. He opened it and thumbed through the pages.

"Scarlett O'Hara," he said. Christopher jotted it down in his Moleskine notebook. Nick took a bite and tossed it on the ground. He told us there were so many out there that he couldn't possibly eat them all, but he said, "I take a taste of every one, every year."

We'd hardly finished the Scarlett O'Hara when we saw an all-white apple with almost translucent skin, which we soon learned was a White Pippin. It was bitter. Nick led us to a Spice Russet that was not particularly attractive — a dull gold with the rough rustlike skin that names these types of apples. Well named, this apple had a surprising taste of cinnamon and nutmeg.

Early the following spring, we ordered mostly standard rootstock and grafting tools. We visited Nick again to pick up 39 varieties of apple scion wood that we grafted onto one hundred roots. Thirteen years later, about half of the spindly tall trees have survived our steep hillside growing conditions, marginal soil, varying degrees of maintenance, and (after our dog, head

of orchard security, died) marauding deer; none of the original Redstreaks made it.

We have not been back to see Nick. In the meantime, Christopher trained with internationally recognized cider authority and educator Peter Mitchell. Children grew. Vegetables, legumes, and grains filled the intervening years, but we always made cider in fall. As we write, Nick is in his nineties and his "ark of apples" orchard is for sale, his legacy destined for new hands.

In 2016, Kirsten was talking to a neighbor who mentioned that he was grafting apple trees. A group of local folks had gone to collect hundreds of varieties in scion wood from Nick's orchard. By then many of the trees had a fungal disease, anthracnose, where infected trees develop dark, water-soaked lesions on stems, leaves, or fruit — common in the moist spring conditions of western Oregon. The group's mission was to find trees that had resisted the attack. The Redstreak was one of them. A few days later, Kirsten sat in the neighbor's greenhouse on a cool, rainy day, cutting twigs and lining up the cambium tissue.

Today there are six Redstreaks and about three dozen other varieties, including wild apples found on the edges of the forest and some that we found on an old mining claim, grafted on semidwarf rootstock in 1-gallon black nursery pots and waiting to be planted in small orchard blocks scattered on our hillsides. Our few dozen apple varieties aren't even 1 percent of what Nick managed for decades by himself and at a much later age, and yet there are times —when we are behind in spring pruning or summer watering or fall harvesting — that it all seems a bit overwhelming to the two of us. In late fall when our will has been parched by the prickly, hot, dry, and now often smoky days of forests of the American West, we wonder if we still have it in us. Then rain settles the sharp edges of the land and the fear of fire, soft green rises through the brown, and we fall in love again. Winter is a time when the farm asks little of us, giving time back to reflect. The question "Is it worth it?" starts to fade deeper into the mist that winds through trees and ridges beyond, and we begin to scheme about things we could plant — raspberries, maybe wine grapes, or grafting more of our favorite wild bittersweet crabapple, adding more pears for perry, or growing out pippins from the seeds of our apples to see what they become.

CIDER:
A SUSTAINABLY MADE, HEALTHY-ISH ADULT BEVERAGE

What if you could help heal our environment in a small way, improve your health, and have fun through the drink? The last one is pretty familiar to everyone, but we don't usually think of drinking as a way to better health or a greener planet. It usually isn't. But in the case of fermented apple cider, it can be all three. It's a sustainable way to improve your health and have fun doing it. To get to that level, you will have to do one more thing: move from being a consumer to becoming a maker.

HEALTHY

Most of us have heard the stories about our colonial forefathers in America who regularly drank local hard cider because it was safe to drink, unlike the water. Even the little ones enjoyed a drink called ciderkin (see page 196) made from the pomace left from apple pressings and the aforementioned unsafe water, which was rendered safe by fermentation. Thankfully most of us have access to clean water these days, but we have plenty of other risks to our health that hard cider can address in a small way. As we talked to cidermakers from around the world, we kept hearing the same thing: "real cider is food."

The vitamins in the fruits used to make cider don't degrade in the fermentation process, and in some cases are enhanced by fermentation. We also get the benefits of the apple's high antioxidants and polyphenols, both of which can help prevent several noncommunicable diseases like diabetes and cardiovascular disease. Hard cider is naturally low in sodium and gluten-free too — at least it is supposed to be (see box on opposite page). Homemade ciders, especially those that are wild fermented, often have lower alcohol levels.

Phytochemicals. These are a group of chemicals found in plants that benefit our health, such as carotenoids, flavonoids, isoflavonoids, and phenolic acid. One of the most important benefits of phytochemicals for us are their antioxidative characteristics. In the United States, apples make up a person's largest source of phenolics.[1] The highest concentrations of phytochemicals are in the apples' peels, which we suppose is yet another reason not to make cider with peeled apples. Phytochemical concentrations vary by varietals, with the top three being Fuji, Red Delicious, and Gala, respectively.[2]

Probiotics. You may have spotted labels on commercial ciders that say "contains probiotics." By most definitions, probiotics are microorganisms that are not only nonpathogenic to humans but have been shown to be beneficial in some way — usually by improving the function of our gastrointestinal tract. A single apple has over 100,000 different microbes,[3] and 90 percent of those are in the core of the apple, which usually goes uneaten but is definitely part of the cider. Included in these microorganisms are lactic acid bacteria (LAB), which are present from apple to finished cider — as long as they aren't killed by the cidermaker. If you want a probiotic cider, you can't apply sulfur dioxide (SO_2) and you can't pasteurize your finished cider, which is probably why it's so rare to see a cider advertising its probiotic content. (That and the fact it represents an additional cost in sending samples to the lab for active colony counts.) When you make your own cider, you can control all these aspects and can produce a probiotic-rich cider.

SUSTAINABLE

There is nothing sustainable about buying bottles of cider if they are made thousands of miles away, shipped, and refrigerated until you feel like picking up a six-pack or a lovely bottle for a special occasion, though it's a far cry from the worst thing you can do to the planet. While you might be supporting a great small-scale cidermaker who is growing apples in a sustainable way and paying staff a living wage, transportation and refrigeration pretty quickly negate the good things. If that cidermaker is local and happy to refill your growler, or they supply your local growler shop, then you

NOT-SO-HEALTHY COMMERCIAL INGREDIENTS

There is a lot of wiggle room in the definition of "cider," so larger commercial or corporate varieties can include a lot of different and unexpected things, which can go a long way toward negating some of the healthier aspects of hard cider. Commercial cidermakers can, for example, use added sugars, including corn syrups, which can drive up the sugar to 12 to 24 grams (3 to 4 teaspoons) or more per bottle. Most of these ciders utilize sulfites, which many people react to, at differing levels of concentration. They can be made from cheap imported apple juice concentrate from Asia. Finally, flavorings and colorings — which can include a wide variety of things — are allowed. So, cheap commercial ciders aren't going to make you healthier and in fact, they are probably taking you the other way. You can change that by making your own cider or by finding a local cidermaker that is making the appropriate decisions to create a healthy product.

are golden. Most of us aren't that lucky, and that means we need to look at ways to get our healthy cider sustainably. Cider is a particularly sustainable beverage to make because it is produced from perennial tree crops that are easy to grow and harvest, and the fermentation process itself doesn't require any large machinery or even electricity (at least at the home or small farm level).

Let's start with raw materials. If you wanted to make your own extremely local beer, you would need to till up the backyard or a local vacant lot to plant your grain crop. Grains take water through the summer, and then there is harvesttime, when you are going to need a combine or else do a lot of hand scything. We speak from experience because one year, Christopher got it into his head that we would produce our own wheat for bread and beer. Since we don't own any big equipment, this meant hand-spreading seeds, then fencing out and diligently defending the young crop from our livestock and the wildlife. Finally came harvest, which became days in the intense late-summer sun to slowly cut one swath at a time, back and forth, back and forth with a scythe. We bundled the harvest, then trucked it to a local thresher for a day of ear-splitting threshing before returning home with one 55-gallon drum of hard-earned wheat berries. Wine is a bit easier because grapes are perennials, but you need trellising and you need to beat the birds to the grapes in fall, when you will need a destemming machine or you will go mad doing it by hand, which we also nearly did.

That brings us to apples. You can fit an apple tree or two from a nursery in the trunk of your car, and they will eventually grow to produce all the eating and drinking apples your family can handle from a corner of your yard. Small trees will even grow in pots on a balcony. They will provide you with shade, absorb carbon dioxide, and sequester carbon, all of which our planet needs.

If planting your own trees isn't feasible, then supporting a local apple grower that is following healthy orchard practices is the next best thing. We have an orchard and understand that when surprise pests or fungus strike it's tempting to turn to a quick chemical solution that promises to make the invaders go away. The problem is that the pesticides, fungicides, and herbicides don't go away, and that can be bad for the environment and for us as consumers of the apples. If you are buying your apples from a grocery store, look for organic apples; this won't guarantee that pesticide residue is not present, but there will be a greater chance that the apples will be free from chemicals you don't want in your cider. If you live in apple-growing country, find a farmer or homeowner who happens to have an old orchard and talk with them about their orchard management practices. If they tell you about field pheromone traps, building the health of their soil and trees, and encouraging beneficial insects through an integrated pest management plan, then you have a good source of apples for your cider. All that said, if you have a local source of apples, even if they are not organic, they can make excellent cider.

Sustainability continues in cider's favor through the making process. No heavy harvesting machinery or destemming equipment is needed.

A hand-operated cider press will last for generations, is simple to operate, and may not require electricity. You don't need power for the fermentation either — just some containers and time. Still or low-carbonated ciders can be bottled safely in recycled glass beer or wine bottles, helping to keep those resources out of the landfill.

Cider can reduce food waste. Have ugly imperfect apples and fruit (organic or not)? Use them in cider. We've met folks from all over the country who make cider with apples that are harvested from abandoned places and empty lots, or yards where the owner is happy to have someone pick and use the fruit. Home cidermaking can be a zero-waste endeavor, as you can recycle the pomace from pressing to make ciderkin or scrap vinegar, feed livestock (even if not your own) or worms, turn it into compost, or use it in heavy concentrations to suppress weeds in your garden. Even the lees (the played-out yeasts from fermenting) can be reused in baked goods or salad dressings, or else composted to become healthy soil amendments in the garden or in potted plants.

If beer is country music and wine is classical, then cider is a playlist of your favorite tunes at the moment.

FUN

There is something for everyone in cidermaking, whether you live in a tiny home, a city loft, or an apartment in an urban or suburban neighborhood, or have a few acres to play with. If you enjoy precise processes, the French traditional method of keeving (page 170) might be just your thing. More of a free spirit? Go capture some wild yeast from spring flowers in a local abandoned meadow, make a culture, and watch it breathe life into a gallon of store-bought apple juice and you will never be the same again. If you enjoy growing things, there is a world of possibilities — from a pot of botanical flavor on your windowsill (think tulsi, white sage, showy hibiscus, fragrant chamomile, or lavender) to other fruits to complement the apple.

Making cider as a community from fruit trees in the neighborhood is a wonderful way to bring people out of their homes and share in the bounty of fruits that are often left to drop. Kids love juicing apples and dipping a cup under the flowing nectar that is fresh-pressed apple juice. Beer makers love the challenge of learning a new fermentation technique, and we have met some who are working on amalgamations of both brews. And everyone loves to share a bottle or two of something they had a hand in making. Anything that brings us together around a table is a good thing in our worldview.

And finally, there is flavor fun. If you love the flavor adventures found when collaborating with microbes and plants, then you'll love the juicy, colorful, and scenic trail of fermenting apples. Yeast, bacteria, enzymes, apple varieties, added fruit and/or botanicals, time, and place all influence what you will pour out of the bottle. The possible aromas, tang, mouthfeel, sweetness, acidity, bouquet, fruitiness, spice, tannin, wood, and fizz that come out of that bottle are endless. It is our hope that the methods and recipes in this book inspire your journey.

WHY THIS BOOK?

There are plenty of good cider books. We know, we have a shelf of at least a dozen of them. So why should you add this one to your shelf? Our cider philosophy is a lot like our farm philosophy, or our philosophy of life for that matter: Learn the basics, then riff on them until you discover something amazing that you truly love. Repeat and always have fun while you are doing it.

If you are someone who loves any or all of these drinks or who thinks that something this good is worth making yourself — or both — this book is for you. Maybe you are interested in going all in and creating a fall family tradition of cider pressing and cidermaking, or maybe you are thinking that infusing wildflowers into some bland hard commercial cider as a way to hack a complexly flavored cider with tastes from your local flavorshed sounds pretty cool. Either way, we want to help you on your journey, whether you need to know the science well enough to understand what's going on and why, or whether you don't care and just want a few recipes to get the idea and go. In short, we want to offer you a foundation to begin or expand your journey with the fermented apple.

If beer is country music and wine is classical, then cider is a playlist of your favorite tunes at the moment. It's an adventure, and it starts with finding a flavor worthy of your quest.

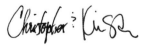

WHAT'S IN A NAME?

In nearly all of the countries in the world, with the exception of the United States and Australia, when someone says *cider* you can presume that they are referring to the fermented, alcoholic version. It was the same here in the United States before prohibition, but as a result of the crusade that did a number on cider consumption and production, we adopted the term *hard cider* for the alcoholic version and *sweet cider* or just *cider* for the pressed fresh juice. In Spain it's *sidra*. In France and Belgium, it's *cidre*. In Mexico and Argentina, *sidra*. In the United Kingdom it's *cyder*. In Austria it's *most*, and in Germany it's *Apfelwein*.

We think it is time for Americans to fall in line with the rest of the world and shake the *hard cider* denomination that has been with us for a century. From this point onward, we will refer to fresh pressed juice as sweet cider, the store-bought pasteurized version as apple juice, and fermented juice as cider, though you can add the *hard* in front of the word each time you read it.

PART ONE

MAKING CIDER

"'When I use a word,' Humpty Dumpty said, in a rather scornful tone, 'it means just what I choose it to mean — neither more nor less.'"

Through the Looking-Glass
by Lewis Carroll

CHAPTER 1

CHOOSE YOUR CIDERMAKING ADVENTURE

Cider is where the innovation is these days. No longer do you need to follow the traditions of faraway makers if that's not your thing. We believe you should make what you love — whether it's bubbly or still, sweet or dry, cloudy or clear — and this chapter will encourage you to do just that.

THE CIDER RENAISSANCE

What was old is new again. Our planet's citizens continue to drink more cider every year, a trend that has continued since the 1990s. How much? In 2017, people all over the world consumed a total of 691,118 gallons (26,161 hectoliters), which in case you are wondering would fill an Olympic-size swimming pool plus about 350 standard bathtubs. Consumption is not equally distributed across geographical regions, however, as you can see from the graphic below. Cider lovers in western Europe consume in one day over six times the amount of cider consumed in the Middle East and North Africa in a year. North Americans have pulled ahead of the Australians and have Africans to pass next, if trends continue. What's most interesting to us is that in countries with traditional cider markets where the process is more strictly controlled or where less flavor innovations exist, the last 5 years' compound growth rates are the lowest. Places like the United Kingdom, France, northern Spain, Germany, and Ireland are all growing their cider consumption at less than 2 percent, while in the United States, Portugal, and many of the former eastern bloc countries, the growth is pegged at double digits.

Place and Process

We think there are really two driving forces at play behind the cider renaissance in the United States: the drive to copy traditional methods, rooted in place, and the desire to innovate new processes.

Some people want to experience and honor the traditional styles of cider in other cultures. That could explain America's recent demand for sour and funky Basque and Asturian Spanish sidras or the light and sweet traditionally made French cidres. Both employ unique processes and produce ciders that reflect that uniqueness. The goal is to re-create that traditional taste, which

Percentage of Worldwide Consumption of Cider*

Figures are for 2017

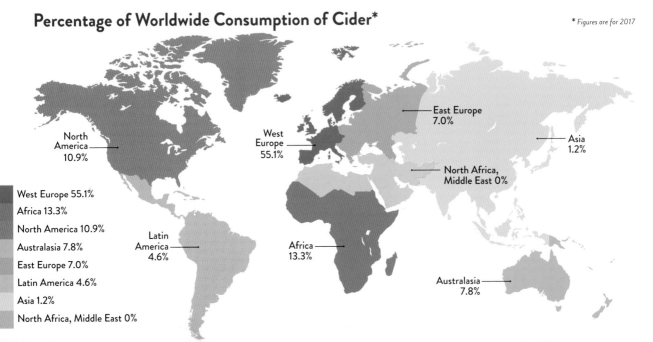

East Europe 7.0%

Asia 1.2%

North America 10.9%

West Europe 55.1%

North Africa, Middle East 0%

Latin America 4.6%

Africa 13.3%

Australasia 7.8%

West Europe 55.1%
Africa 13.3%
North America 10.9%
Australasia 7.8%
East Europe 7.0%
Latin America 4.6%
Asia 1.2%
North Africa, Middle East 0%

can mean everything from planting the same trees to using the same yeasts to building replicas of the same processing equipment to using the same types of bottles. It's about re-creating the same paints, the same types of canvases that the masters used, and then copying their style. It is a quest for that heritage, that traditional taste.

The other driving force honors the first, but then goes in a completely innovative direction. This is about creating something new. It's about mashing up flavors, processes, yeasts, bacteria, and fruits and botanicals (yes, even cannabis) to create something fresh and different. We go from specific apple varietals to all fruits, from commercial strains of wine yeasts to harvesting wild yeasts from what's blooming in your own backyard right now. It's about moving from the 8-crayon box to the jumbo 120-crayon box and then coloring well outside the lines. It is a quest for something delightfully original.

NO JUDGMENTS HERE

Spend any time with cidermakers and you will realize they are an opinionated bunch who aren't afraid to share with you their views on things that matter to them, like, say, what is a real cider. For some, it's one made with heritage apples or pears (in the case of perry). For others, it's only those made with wild yeasts or only blessed commercial yeasts. There are even disagreements over the humble bubble, as some are adamant that a real cider is finished dry and still, while others don't mind forcing a little carbon dioxide into the bottle for a nice pop. Some believe you must age a cider

Cider made with the tree lichen usnea

to proper maturity in a bottle or an oak barrel while others are fine with a quickly fermented, forced-carbonated cider that's canned and out the door of the cidery in less than 2 weeks. The situation reminds us a lot of politics — no matter what you believe, you can find a tribe of people who are just as passionate about that as you are, and if you hang around those people all the time it can affect how you see the world and those with other beliefs.

Perhaps the one thing that will unite all of these opinionated cider people is the word *vinegar*. To a cidermaker, the vinegar-creating bacteria acetobacter is like public enemy number one — nothing less than locking it away or complete extermination will do. The fear is that it will "infect" their cider, ruining it and future batches. This means that vinegar and cider are like two relatives that refuse to mention the other in polite conversation. You have to find the right moment, maybe after an afternoon cocktail or two, to even broach the subject of the other being related. Talk to craft vinegar enthusiasts, however, and sometimes you have to remind them that yes, their health-giving acidic tonic was once a naughty alcoholic version of itself. It often goes to a "let's not talk about alcohol, let's just talk about our vinegar" kind of place.

To us, all are welcome. Yes, we understand the arguments. We built his and hers fermenting caves into our house (Kirsten's for all things sour, Christopher's for cider), after all, and for more years than we want in print we have been making

our ciders and vinegars while keeping a careful eye on the other guy's operation. Kirsten has been filching Christopher's cider to make her vinegar and Christopher will now publicly admit that she can take one of his average ciders and turn it into an amazing vinegar. Looking at this holistically, it is all part of one natural process that happens with or without humans. And to be fair, humans drank alcohol for ceremony or to avoid questionable water, but they also used that next stage in the fermentation — vinegar — to preserve harvests and heal bodies. Without both of these permutations of sugar, it is hard to say where humans would be now. So why not allow that all are good and the only choice you need to make is not which side to be on, but what do you want to make next?

We intended this book to include a chapter on vinegar, both for those who start with the intention of making vinegar and those who want to salvage a batch of cider that doesn't turn out the way they wanted. Then Kirsten went down the rabbit hole, and when she emerged after a year of vinegar research and countless jars and crocks of vinegar trials, she had a huge chapter. The techniques and recipes were exciting — pushing the edges of what had been done before — but the chapter was larger than any other chapter in this book. That is how our forthcoming vinegar book was born, by being booted from this book. For those interested in vinegar, we think it makes a great companion book to this one.

Ten-year aged cider vinegars

MAKE WHAT YOU LIKE

Our motto when it comes to making cider is "Life is too short, so make what you like," which means you need to know what you like before you can make it. That doesn't necessarily mean you'll be able to reproduce what you find in your refrigerator door or what you order at the bar on a night out with friends — but knowing what you like about those beverages will help you craft something that makes you happy.

We met Jordan Werner Barry when her Food52 *Burnt Toast* team interviewed us about bubbles, of all things. While vegetable fermentation has some pretty impressive bacteria-produced carbon dioxide, it's honestly nothing like what yeast can do in a few gallons of fresh and sweet fruit juice. After we traversed the topic of bubbles as part of the postsession wrap-up, we learned of Jordan's 2 years of work cataloging the language of cider producers from around the United States as part of her master of arts degree from NYU. You can learn more about the project at her website Cider Language (see Resources, page 321), but our takeaway was that makers and consumers don't agree about what simple words like *dry* or *fruity* mean. Even the word *cider* means different things to different people, so you can be forgiven if it all seems a bit muddied. In the rest of this chapter, we are going to try and break all of this down into a few simple questions, and based upon your answers, give you some recipe recommendations to start your adventure.

How Much Alcohol?

If you are already a cider enthusiast, pick your three favorite commercial ciders and find the alcohol by volume (ABV) percentage on their labels. That should give you a good idea as to the range of alcohol you enjoy.

If you are a beer or wine drinker new to cider, the good news is that you can make a cider that has the alcohol content that you already enjoy in your favorite drink. In general, most beers range from 4 to 8 percent ABV, though you will find low-alcohol beers below that and some special craft beers now pushing 12 percent! On the wine side of the world, something like a Moscato d'Asti starts things off in the 5 percent ABV range while some heavy reds, like some California Zinfandels or sweet dessert wines like a Muscat, can hit or exceed 15 percent. To get above that level, wines need to be fortified with booster rockets of distilled spirits, as found in port, Madeira, and sherry. Well, it's the same game for cider.

RECOMMENDATIONS

Here are some recommendations for ABV percentages, so you can start to think about what you want. We'll cover how to get these specific ABV percentages starting on page 78.

LESS THAN 6 PERCENT ABV. There are a lot of good reasons to prefer a lower level of alcohol in your cider. You might have heard of the term *session beer* or *cider* — basically the concept is that during a drinking session your drink of choice should have a low enough alcohol level to not impair you at the conclusion of said session.

Active primary fermentation of cider with raspberries

may cause the cider to spoil if it has an ABV of less than 6 percent and it is not pasteurized. For example, when we talked to sidra makers in the Basque Country of Spain, they explained that traditionally, the sidra was mostly consumed by sometime in May, because as the high temperatures of the summer months warmed the barrels, the bacteria awakened, giving these ciders their funky reputation. Our suggestion is to store these ciders in your refrigerator and enjoy them within a couple of months.

To achieve this low an ABV, you might need to blend a lower-sugar fruit like guava, raspberry, or gooseberry with your apples or dilute your apple juice with unchlorinated water to bring down the sugar levels. A third option is to rely on yeasts that are alcoholic lightweights (we'll explain later).

6 TO 8 PERCENT ABV. This is the sweet spot for most apple varieties, as well as other fruits that you might want to blend into your cider. It is

Think of a favorite tailgate party or keeping a few bottles in the creek while you are mending fences all afternoon. When it comes to cider, besides being lighter-drinking, the upside is that, when unpasteurized, it is likely to be the most probiotic. That said, those same live bacteria

---

ABV:
ALCOHOL BY VOLUME OR APPROXIMATE BUT VARIABLE

One caveat that cidermakers already know: ABV percentage on the label isn't necessarily correct. That's because there is some wiggle room in the law to accommodate small craft cidermakers who have some variability in their batches. The actual ABV doesn't vary by much — less than 2 percent — but you should be aware of it. While tracking the sugars through fermentation gives the cidermaker a good idea of the final ABV percentage, it's not perfect. To know exactly, you would need the help of a lab, although you can buy photometric kits that promise to make this process far easier and less expensive than sending samples to a certified lab or building one in your cidery. Still, we suggest considering the ABV percentage on your favorite labels as a close approximation but not an absolute.

cider's natural range, above many beers but below many wines. It might be too high for a bottle-pounding "session," but it's light enough to drink responsibly and still find your way home.

Making a cider in this alcohol range is pretty easy. In our experience, both purchased apples and unpreserved juice will have a sugar level that will land within this range, and nearly all the commercial yeasts will happily ferment this amount of sugar to full dryness.

ABOVE 8 PERCENT ABV. Just like relatively low-alcohol "session ciders" have a function, so do the ones at the other end of the ABV scale. Maybe you are a wine drinker and you would like to produce a bottle in that typical 12 to 15 percent ABV range of most wines. You can extend your reach even further into ice cider and pommeau for a nice heavy sipper at the end of the day. How high you can go has everything to do with the alcohol tolerance of your yeast (you'll need to use commercial yeast) and keeping them fed.

Still or Bubbly?

Most ciders finish still — or what some describe less flatteringly as flat. That's because the carbon dioxide produced by the yeasts — the bubbles — escapes out the top of the carboy through a special piece of equipment called an airlock, which is designed to allow the carbon dioxide to escape without letting outside air into the cider. If you want still cider, transfer to your bottle of choice, tuck away for a few months, and you are set.

If you want a satisfying pop when you open that bottle, it just takes a few extra steps to trap

The same cider recipe with and without carbonation

some carbon dioxide in the cider in the bottle. There are several methods to choose from, and with practice, you can actually refine the process to the point where you can choose the type of bubbles you want. Seriously. You will read more about this in chapter 3.

Dry or Sweet?

This is probably the most misunderstood and maligned area of cider. You don't need to look far to come across cider folks railing against the scourge of sugary sodalike ciders (likened to wine coolers) that have become so common on store shelves. These are purists who flatly state cider is meant to be dry, end of discussion. As surveys from several cider societies in North America have found, people even think they like dry cider when in fact they are choosing again and again semi-sweet or even sweet finished ciders, and the confusion is not necessarily their fault. The word

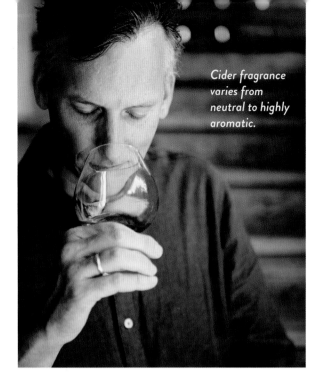

Cider fragrance varies from neutral to highly aromatic.

Ciders finish from clear to cloudy, depending on a number of factors.

dry is the most common word found on American commercial cider labels and not all of these are, in fact, dry ciders.

This is a no-shame book. Our goal is for you to make what you love, remember? If you love something bubbly and sweet, then that's what we are going to make together, and you will love it. Making a semisweet or sweet cider just takes a little more work to make sure some sugars are left for us to enjoy, which we will cover in chapter 3.

Neutral or Fragrant?

Aroma is something we may not even think about because some large-scale commercial ciders don't have much to offer up to our noses before our lips turn. It's too bad, because cider is a great drink to showcase all kinds of botanical aromas that are released when the cider is poured in the glass. Transported up to your nose by thousands of rising and busting bubbles, aroma is really the first opportunity you have to "taste" your cider.

If you haven't experienced the delightful fragrance of cider, look for a well-made commercial one made with hops, which are much loved in many IPA-style beers. Pour a glass and hover your nose just above the bubble-bursting surface and breathe deeply. Can you see those little green-feathered hops hanging from their trellises? They really set the stage for what your mouth will experience next, yet the flavor doesn't have to be as bold as a hopped cider. What about spring flowers from your yard? Blossoms from old fruit trees in the neighborhood or alongside a back-country road? Spices or "adjuncts" like dried orange, lemongrass, or cocoa husks? Everything is game, and once you start playing with this layer you can bump up the complexity of your ciders. Throughout the book you will find recipes that will help you discover new cider aromas.

A Clear or Cloudy Day?

Think about what you see when you pour cider. Some ciders are crystal clear — so clear, in fact, that you can read a menu held on the opposite side of the cider glass, which might come in handy if you are out with friends and have forgotten your reading glasses. These ciders really show off the amber spectrum of colors possible with ciders. Add other fruits like berries, plums, and cherries, and you have a bigger palette to paint from. While some ciders clear right up, others don't necessarily become crystal clear naturally, so it often takes some finesse to understand the early conditions and if you will need to make interventions, like adding pectic enzymes (see Clearing Things Up with Pectic Enzymes on page 77 for more on this technique).

But let's not forget how enjoyable a cider that is a little cloudy can be. Many styles, like scrumpy and farmhouse, are typically a little on the cloudy side, due to the level of fruit solids and yeasts. For those of you seeking that healthy cider, these bits can carry with them some probiotics and a little nutrition. It can be very enjoyable to peer into your cider and not see the menu on the other side. It's kind of mysterious, like what's in there? As you might guess, the trick to making a cloudy cider is not to intervene.

Do You Want to Get Funky?

There was a time when the only ciders that were funky were some European styles or home-brewed ciders offered up with embarrassment. Years ago, soon after we started making ciders on our farm, we started getting some batches that were different from everything that had come before — really different. Not bad, certainly drinkable, but nowhere near something we could get at the store. We chalked it up to black magic and prayed to the cider house gods to not let it happen to all of our ciders.

Fast-forward more than a decade, and funky is hip. What used to be relegated to the faults section in cidermaking classes is now sought after by some. Now we get at least one person in each of our cidermaking classes who raises her hand and asks how to get a little funk in their ciders. The good news is that it turns out to not be black magic after all, and with patience and practice you can easily add just a bit of funk if you want (or know what to do next time if you don't want it but find it creeping into your batches — see page 307).

It's All Up to You

Remember our motto: "Life is too short, so make what you like"? It really is, and this is the point in your journey where, having looked at the maps, asked passersby what's interesting in the area, and gauged the weather, you have to pick a direction. The good news is there is no bad choice and with time you can explore everything — even cider places you had no idea existed or maybe you had heard of them but could never imagine yourself going there. Maybe after a few great batches that you are quite proud of, you decide to go adventuring that way and you love it. That's our hope and why we wrote this book. Okay, now that we are all thirsty, let's get to making some cider!

CHAPTER 2
EQUIPMENT OVERVIEW

To make cider you only need a few key pieces of equipment: a container with a lid to contain the fermenting sweet cider and bottles with caps to store it once it's fermented and before you drink it. Everything else helps with the process, either by making it safer or by making it easier on you as the cidermaker. We will suggest three setups, from the simplest to get you started cheaply to the sublime, in case you want to go for a perfect setup.

APPLES TO POMACE TO JUICE

If you are going to press apples, you may or may not need any special equipment — depending on how much cider you want to make. Making six gallons of fresh-pressed apple cider requires about 80 pounds of apples or around two common apple boxes. You can pick a couple of boxes' worth of apples pretty quickly, especially if the trees are on dwarf rootstock and easy to reach. You can also purchase a couple of boxes of different varieties of organic apples (see the next chapter for more on how to decide which varieties to buy). At this scale you really don't need any special picking equipment. Before investing in juicing equipment, check your local brew shop to see if you can rent a grinder (sometimes called a mill) and press, then invite your friends over to help.

There are many options for grinders and presses (see Suppliers, page 321). Usually they are paired, meaning a press will include a hopper that feeds the apples through a grinder/mill and into the basket or racks. The biggest thing to consider is the size of the chop of the pomace — it should match the ideal size for your press. If the chop is too big, you won't get as much juice as you should; if the chop is too fine, it can gunk up your press, leading to frustration and lots of extra cleaning of the cloths between loads. If you are handy in the shop and are looking for a project, you can find a whole host of instructions out there for building your own mill and press, though we believe the most complete reference is Claude Jolicoeur's excellent *The New Cider Maker's Handbook*, which devotes a chapter each to mill and press designs.

If you want to make more than 6 gallons, you are going to want to pick your own apples and you will need to borrow the mill and press — or you may become addicted to the fun, in which case it's time to think about purchasing one of your own. Apples don't all ripen at once, so if you are lucky, this will be more manageable as a series of afternoon pressings as the apples ripen.

FERMENTATION SETUPS

Turning fresh, sweet cider into its next naturally occurring alcoholic phase requires the help of yeasts, time, and containers. There is also a choose-your-own-adventure aspect to cidermaking equipment, depending on how deeply you want to dive in. We present three levels of setup, from minimal to majestic. You may wish to start at the minimal setup and then just add more equipment as your familiarity with the process grows.

JUICY AND NOT-SO-JUICY

Be aware that in our experience, different apple varieties can yield differing amounts of juice using the same press. When we have a basket that doesn't produce as much juice as we expected, we do one of two things: we either leave it under pressure for twice as long as usual, which for our setup is about 15 minutes, or, if we have a lot of apples to get through and don't want to delay the process, we reserve the pressed pomace in 5-gallon food-grade plastic buckets and once the rush of the apples is finished, we run the pomace through again at the end of the run with a long time under the press. Often this waiting period and second press will give us the juice we were missing. This happens every year, both when we're pressing apples from our orchard and when we're pressing purchased boxes of single varietals for recipe testing during spring and early summer when our farm harvest is long gone.

EQUIPPING YOUR CIDER HOUSE

LEVEL 1: Minimal Setup

- Food-grade buckets with lids
- Airlocks (optional)
- Funnel
- Bail-top-style bottles (new or used)
- Commercial yeast (optional)

LEVEL 2: Simple Setup—
All of the Above PLUS:

- No-rinse brewing sanitizer
- Hydrometer
- Wine thief (basically a glorified turkey baster)
- Plastic graduated cylinder to hold hydrometer (optional)
- Racking cane (autosiphon models are easier to use and more sanitary)
- 6–8 feet food-grade vinyl tubing
- Glass or plastic carboys
- Thermometer

LEVEL 3: One Fine Setup—
All of the Above PLUS:

- Floating thermometer
- pH strips or handheld pH meter
- Brass V-shaped carboy and bottle washer
- Bottle-drying tree
- Hand capper or corker
- New caps and corks
- 6-gallon FerMonster plastic carboy
- Cornical fermenter
- Pectic enzyme powder
- Campden tables (optional)
- Oak chips, spirals, or sticks
- Recycled or new food-grade barrels

WANT TO EXPAND? CONSIDER EQUIPMENT UPKEEP

Christopher is not the handiest guy when it comes to all things mechanical. His style of equipment maintenance is more like patching or replacing things when they leak too much. This is honestly why we didn't go into commercial cidermaking. After we realized the amount of used equipment that we would need to purchase, and more importantly that Christopher would need to maintain, we made an honest assessment of skills and passion and decided to keep our professional fermenting to vegetables. Making sauerkraut requires far fewer pumps, hydraulic lines, and filtering systems. Zero as a matter of fact, at our scale. So, if you are considering a micro-cidery of your own, think about who will keep everything running in tip-top shape.

Level 1: Minimal Setup

Starting at the most basic — and cheapest — level, you need a food-grade container to hold the juice as it ferments, a lid or (better) an airlock to keep air away from your fermenting cider, and a funnel and bottles. Everything can be washed with dish soap and warm water and rinsed and dried completely. Lastly, unless you are making a wild yeast cider, you will need a commercial yeast to get things going. That's honestly it.

With this setup, you won't be transferring the cider from the primary to secondary fermentation (racking) but rather letting it ferment to the desired sweetness/dryness level and then bottling. You will need bottles. An inexpensive option is to reuse quality beer, cider, and wine bottles. If you reuse the bail-top-style bottles you won't need anything else. For all other bottles, you will need either a hand capper or a hand corker, depending upon which style you are filling. New bottle caps and corks are available in different colors and sizes, so make sure you match your bottles to the right size cap or cork.

Bottling will be a bit trickier because without a racking cane and hose siphon to control where you pull off the cider, you will need to tip the bucket and carefully pour off the cider above the lees into the funnel and bottles. You'll need a steady hand and a lot of patience to pull this off well. This setup will work but probably not every time. Adding a little more equipment can improve those odds.

Level 2: Simple Setup

By adding a couple of pieces of equipment to the minimal setup, you can make your life easier and improve the hygiene of your cidermaking. Starting with hygiene, improve your chances of removing any unwanted bacteria from your equipment by using a no-rinse brewing sanitizer. You simply mix the sanitizer per instructions and either submerge your equipment into the sanitizing solution or rinse your equipment with it. The racking cane and tube need to have the sanitizing solution run through them and be allowed to dry. As the name suggests, you don't even need to rinse off the sanitizer afterward. See Hygiene (page 84) for some suggestions based on what we use.

To make your life simpler we recommend adding three pieces of equipment to the minimal setup: a hydrometer, a racking cane attached to a 6- to 8-foot length of food-grade vinyl tubing, and a wine thief. With the hydrometer, you can measure the amount of sugars in your juice, which is helpful to know when you are aiming for a specific alcohol content and sweetness. You might also spring for a plastic graduated cylinder, which

Floating hydrometer to measure specific gravity

Extracting cider with a wine thief

holds the juice or cider in a vertical tube so that when floating your hydrometer, you can clearly see the readings. We broke our first one right away and tried unsuccessfully to use a number of other glass jars as substitutes, so best get this now and treat the hydrometer and cylinder gently. Along with measuring specific gravity (SG), it's also useful to measure the temperature of both the cider and the proofing liquid of the yeast. It's important to know these temperatures because you don't want too much of a difference, which may cause the yeast to go into shock and die off prematurely. In that same vein, it's helpful to have a thermometer in your fermenting space

MICRO-CIDER HOUSE:
FERMENTING VESSEL TIP

The 1-gallon glass bottles that hold some commercial apple juice are perfect for small-scale fermenting. They fit nicely on your counter. We recommend you also get some ¾-gallon bottles that are now becoming a common juice bottle size, as they are the perfect size for racking into from the 1-gallon bottle after primary fermentation.

to monitor the fermentation; this is especially important in some processes that require a cold fermentation to retain sugars, like ice cider (page 260) and keeving (page 170). Inexpensive stick-on strip thermometers can be affixed to your fermentation vessels and will give a rough idea of the ambient temperature the ferment is experiencing.

If you have been making your cider in food-grade buckets, we recommend moving up to glass or plastic carboys for two reasons. First, you can see what's going on in your cider without having to open it up and draw out a sample. Second, carboys have necks and buckets don't, which means the area above your cider's surface is wide in a bucket but narrow in a carboy. That's a good thing because it's much easier to manage this space to keep unwanted aerobic microbes out. To move cider between buckets, carboys, and bottles, nothing beats a racking cane and hose. The simplest models depend upon you to create the suction, while self-siphoning models only take a couple of quick pumps to get the siphon going.

Finally, a wine thief is a groovy tool to have. It will allow you to sneak out a bit of cider to check specific gravity and taste what is happening without making much disturbance. As we've grown as cidermakers, we've used this more and more to taste and learn how things are developing. Doing this has allowed us to make small changes as needed and control some aging styles, such as bâtonnage (which you will read about on page 104.)

Level 3: One Fine Setup

The simple setup gets you there, but adding more equipment can make the process easier and allow you more flexibility in terms of what you are making and how you are bottling it. Adding a floating thermometer will allow you to manage your cider's temperature, which is important when introducing commercial yeasts and during the fermentation stages. Adding pH paper strips or a meter will allow you to quickly check the pH of your cider, from fresh juice to finished cider.

Using recycled bottles is a great idea and requires good cleaning, and new bail-top-style bottles can get expensive and you will want to reuse them. One of our favorite yet simplest pieces of equipment is a brass V-shaped carboy and bottle washer that screws onto the threaded end of a kitchen sink faucet. You simply invert the bottle or carboy and ease it down onto the upwardly protruding end of this thing. A stream of water is shot upward into the carboy or bottle and does a great job of swishing everything out.

After cleaning, you need a place for your bottles to air-dry upside down. Bottle-drying trees look a little like the trunk of Christopher's family's plastic Christmas tree from the 1970s, with short nibs sprouting out in rings upward. However silly they may look, they do work and can safely hold a surprising number of bottles.

In terms of fermenters, we love two additions to our glass carboys: a 6-gallon FerMonster plastic carboy and a Blichmann Cornical fermenter. The plastic carboy is much lighter than the glass and has a wide lid, which is the bomb when you are

fermenting with fruit mashes or whole hops —
they go in the narrow glass carboy necks much
better than they go out.

Our biggest splurge in the cider cave has been
a stainless-steel conical fermenter. After reading
for months about the virtues of conical fermentation devices, and after Christopher convinced
himself he couldn't call himself a proper
cidermaker without at least one corny keg lying
about, we bought a Blichmann Cornical fermenter
in a moment of weakness. Basically, it's a corny
keg with legs and a conical attachment, so that
you can ferment and serve in one keg. We have
found the conical design to be great for removing
lees without racking. And the keg component is
great when you find yourself in a situation where,
because of your cider reputation, you have been
designated chief cider supplier for that tailgate
party of 20 or more.

To ramp up quantities, you need to move to
larger containers, and one of the most economical
ways to do that is with recycled food-grade barrels
made of high-density polyethylene (HDPE).
These are commonly sold in 150-liter (40-gallon)
sizes and are known to many wine, beer, and
cidermakers as "blue oak" because they function
as oak barrels, but without the weight or cost.

Lastly, to get the barrel-aged taste that none
of the above will provide, you need some wood.
In the past this meant a wooden barrel, likely
used from a distillery or winery, but the size can
be daunting for a home cidermaker. Smaller new
barrels are available online, made from various
woods. Realistically, smaller than 10 liters

APPLES TO JUICE TO CIDER: WHAT CHANGES?

Surprisingly little changes when you go from an
apple to fermented cider. These eight elements
remain the same in both apples and fresh juice:

- Water
- Sugar
- Sugar alcohols
- Acid
- Nitrogenous compounds
- Tannins
- Aroma compounds
- Pigments

One item that might have caught your eye is
sugar alcohols. You thought your apple or fresh
juice was alcohol-free? Well it is; the name is
confusing and doesn't actually refer to alcohol
or ethanol but to its chemical structure, which is
different from sugar. When we ferment the juice,
alcohol, or more precisely ethanol, joins the list.
If the cider is finished to full dryness, all the sugars are removed, so that one drops from the list.

(2.5 gallons) can be difficult, as we have found
that in our drier climate they tend to dehydrate
quickly. They are definitely cute, but they can be
expensive. There are great alternatives now in the
form of oak chips, spirals, and sticks, and even
something called inner staves — wine-soaked
stave pieces from previous working barrels that
you submerge in the cider as it is aging, which
imparts varying levels of barrel-aged oak taste.

JUICING APPLES WITHOUT A PRESS

For each of these options, rinse your apples and keep the skin on.

Juicers. Each juicer is different, but they are good at what they do: extracting juice. The downside is that they take time. We are patient enough for about a gallon of juice, which takes roughly 20 pounds of apples. In our experience, because the apples are ground further, the juice is much thicker and there is no clarity. We tried treating some batches of cider with pectic enzymes to clarify the juice. The enzymes did their job quickly, but the results didn't seem worth the effort or the juice we lost from having to rack the batch after the enzymes were added. The untreated juice clarified,

making a nice drink that ended up with more flavor; it just took a little longer.

Hand grater. It might be difficult to imagine, but if you want to juice a smallish number of apples and don't have a juicer, this is your next best option. Grate whole apples in a bowl, rotating the apples as you grate to avoid grating the core. It goes surprisingly quickly. You can of course use the grater attachment to a food processor, which is less of an arm workout, but we found that by the time you cut out the core and slice it to fit down in the hopper, and take apart the bowl to empty it periodically, it isn't any faster. Place the grated pieces in cheesecloth and squeeze out the juice, or use a tabletop press.

Blenders or food processors. Most people have either a blender or a food processor, or both, but from our experience they are not the best choices for juicing. The apples can quickly be chopped too finely in a food processor, and trying to get the juice out of the pomace through cheesecloth only makes a mess. Blenders can purée the fruit too quickly. Still, if you'd like to try using your food processor, you can make it work by prechopping the apples into an appropriate chunk size for the processor, then give them a quick couple of pulses to reduce their size for the press. Place the food processor as close as you can to the press, so it's easy to transfer the apples. The juice may have some solids, so it's a good idea to line a large funnel with cheesecloth and use it to funnel the fresh cider from the press into your clean plastic gallon jugs.

TURNING YOUR FRUIT INTO A COMMUNITY CIDER

Cider presses are an investment, and they take up quite a bit of room during the 9 to 11 months of the year that aren't apple season. One way to ease the financial burden and create a sense of community, which is often lost in our modern lifestyle, is to find a group of friends to split the cost and storage of a press, then have everyone come out for a pressing party to celebrate the season. In fact, we first pressed apples into cider in this way long before we moved to our smallholding.

It was a cool, gray drizzling day when we arrived with our boxes of apples, a potluck dish, and passel of small boys, at the home of an older couple who'd done community pressings for years. Soon the grinder was macerating, the press squeezing out the sweet amber nectar of fall. The children interrupted their games to hold cups under the press and the veterans of these pressings told

stories; we'd found a magic that would become part of our own fall tradition for years to come.

If you have apple trees, you might want to check with your local cidermakers to see if they hold community pressing events. Other cideries around the country, like WildCraft Cider (page 230) and Apple Outlaw (page 186), press the mishmash of the communities' fruit to become juice and then cider. You may not reap every drop of your fruit, but you will likely walk away with some fresh juice or vouchers for the resultant cider — not bad for fruit that may otherwise have ended up on the ground to rot. Still, for many people the satisfaction comes in being part of the effort, part of creating that flavor of place. Apple Outlaw takes the zero-waste approach one step further. Any profits from the juice go to the community food bank and other local organizations focused on feeding the people.

The ripe, the golden month has come again, and in Virginia the chinkapins are falling. Frost sharps the middle music of the seasons, and all things living on the earth turn home again... the fields are cut, the granaries are full, the bins are loaded to the brim with fatness, and from the cider-press the rich brown oozings of the York Imperials run."

— Thomas Wolfe, *Of Time and the River: A Legend of Man's Hunger in His Youth*

CHAPTER 3
THE MASTER PROCESS

Here is the secret: the dozens of cider recipes in this book all follow the basic same steps. The biggest difference between them is whether you are starting with freshly picked apples, freshly pressed sweet cider, or pasteurized store-bought juice. Knowing and mastering this process will allow you the flexibility to take advantage of opportunities, be it boxes of apples or peach seconds that need to be used today, or to make a batch in early summer when your only option for juice is likely at the supermarket in a gallon jug.

CIDERMAKING BASICS AT-A-GLANCE

1. Choose good apples.

2. Sweat apples.

3. Wash, then grind apples.

4. Press apples.

5. Blend juices.

GOLDEN RUSSET
SG 1.075 pH 3.2

GOLDEN DELICIOUS
SG 1.070 pH 3.9

LIBERTY
SG 1.059 pH 3.5

6. Pitch the yeast (optional).

7. Ferment.

8. Rack.

9. Ferment some more.

10. Stop the fermentation (optional).

11. Age (optional).

12. Bottle.

13. Drink.

PICKING FRUIT

This process starts with fruit on a tree. Even if you are starting with juice from a local orchard or a gallon bottle bought off a store shelf, it all began as fruit hanging from a tree. The quality of what you drink starts here. Yes, you can make a good-tasting cider from decent store-bought juice, but it's a lot harder to make a great, memorable drink from the same juice.

Memorable ciders are usually based on a balance of the three flavor categories of apples: sweet, sharp (sour), and bitter. Interestingly, these are three of our five tastes, with saltiness and umami rounding them out. Saltiness has never shown up for us in a flavor profile of any cider but umami has, with the help of yeasts in a process called bâtonnage (we'll cover this on page 104). Apples with high sugars and mid-levels of acidity are magical for cider. Kingston Black is an example of an I've-got-it-all apple that can be made into a memorable single-varietal cider. Pink Lady is an easily obtained dessert apple that has enough other flavor qualities to make a decent single-variety cider. Usually, though, you will find this balance by picking several varieties of apples, each of which will be higher in one or two of the flavor categories but usually not all three. Most apples that are grown for fresh eating are low in bitterness, so finding apples with that characteristic is usually the biggest challenge to a cidermaker.

If you have the opportunity to start your cider adventure here, you are lucky. You're in for a remarkable afternoon because there is nothing quite like being in an orchard on a fall day among the trees. That said, there are plenty of amateur cidermakers who start with sweet cider from a local orchard or supermarket (see box on page 74), and this may be the best way to go if you're new to the process and want to dive straight into fermenting. In that case, skip to blending (page 69).

If you are planning to make your cider using apples you pick yourself, make sure they are ripe. How can you tell if an apple is ripe? There are a few simple tricks to know when the tree and the apple have decided it's time to part ways. First, the stem will tell you when it's ready to let the apple go. Gently grab the apple from underneath, lift upward, and twist, like you are screwing a lightbulb into a ceiling outlet. If it releases, it's

Wickson Crab has a uniquely high sugar level with some nice acids and tannins. It will make a high-alcohol cider (9.5 percent ABV) all by itself.

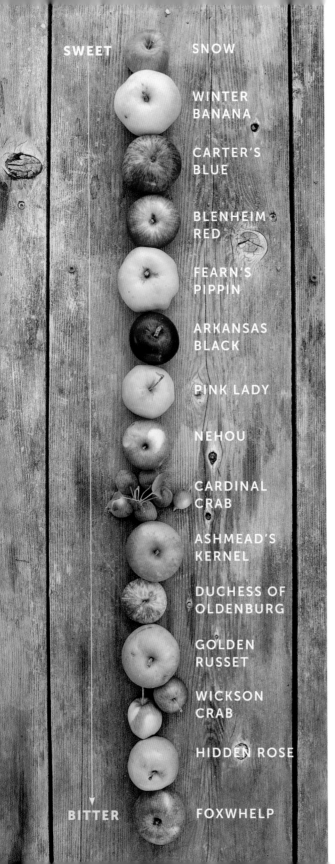

SWEET

SNOW

WINTER
BANANA

CARTER'S
BLUE

BLENHEIM
RED

FEARN'S
PIPPIN

ARKANSAS
BLACK

PINK LADY

NEHOU

CARDINAL
CRAB

ASHMEAD'S
KERNEL

DUCHESS OF
OLDENBURG

GOLDEN
RUSSET

WICKSON
CRAB

HIDDEN ROSE

BITTER

FOXWHELP

ripe. If you have to keep twisting and tugging, it's not ripe. Second, the seeds will tell you. Cut the apple through the middle line, as if your apple was wearing a belt to hold up its pants. Inside you should see the seeds in an inner star and they should be brown, not tan or greenish white.

Finally, there is a simple test you can use to measure how much starch in the apple has been converted to sugar. Iodine reacts to starches but not to sugars, so a few drops of iodine on that freshly cut apple will tell you a lot. If it stays iodine brown, its starches are now sugars just waiting for you. If it turns color — more of a blue, black, or gray — it still contains starches. As the apple ripens, the starch disappears from the center and moves outward, so if you see a small amount of coloring along the outside next to the skin, that's just fine. If there is an indication of starch across the surface, this might not be a problem if you plan on sweating the apples (see Sweating Apples on page 59), because the apples will continue to convert starches to sugar after they have been picked.

You will notice we haven't talked about gathering apples from the ground, and that's because we are not fans of this at all. Have you heard the saying one rotten apple spoils the barrel? The same goes for cider, although it would be more accurate to say that one rotten apple in a bushel basket spoils your cider. *Spoil* is a strong word here, but it can affect the flavor and that can go beyond just an "earthy" taste. Yes, it is hard to see perfectly good-looking apples just lying there and not pick them up, and it's a lot easier to fill your box by scooping them up than it is to climb on a ladder. You may wonder that if humans have been eating apples off the ground (and making cider with them) for thousands of years, what's the big deal?

SWEET, SHARP, OR BITTER?

An apple variety will often have a commercial name, and if it's used in cidermaking, it will have another name that identifies which one or two of the three flavor categories (sweet, sharp, or bitter) it best represents. An apple that's only high in acidity is a sharp, one that is only high in sugar is a sweet, and one that is just high in tannins (bitter) is aptly referred to as a spitter (believe us, they deserve this label). An apple that is high in sugar and acidity is sweetsharp. An apple that is high in tannins and acidity is bitter-sharp. Apples that are high in sugar and tannins are bittersweets. We have a couple of high-tannin cider apples and pears in our orchard, and it is great fun to watch guests begin grazing the orchard knowing they will soon arrive at a spitter tree. Kirsten always warns people; Christopher rarely does.

You really only hear these names in the cider-making world. More likely, you will hear apple varieties labeled as culinary, dessert, cider, and maybe keeper. Let's look at these with our new flavor categories.

SWEET APPLES are often called "eating apples" or "dessert apples." Remember that all of these sugars will naturally be converted to alcohol through fermentation, so picking sweet apples to make cider has more to do with the final alcohol by volume (ABV) than sweetness.

SHARP / SOUR / ACIDIC APPLES are often referred to as "cooking apples" or "culinary apples," and sometimes people get really specific and call these apples "pie apples" because acidic apples typically stand up well under the crust of a good pie, and when combined with sugar and cinnamon, give you the sweet-tart taste an apple pie should have.

BITTER APPLES are often referred to as cider apples. The tannins are likely to be on the higher side, meaning the apples are bittersharps, bittersweets, or our beloved spitters.

Finally, you might hear of an apple variety being referred to as a "keeper." This doesn't actually refer to its flavor, although in our experience it's likely quite tart, but a combination of its natural wax, thick skin, and firm flesh that allows it to "keep" in storage longer than other apple varieties.

The big deal is a specific strain of *Escherichia coli* (E. coli O157:H7) that has acquired specific virulence genes that turn them into pathogens that can make us sick, and in the worst case, kill us. Look, we are not alarmist types or we wouldn't spend so much of our waking lives harnessing microbes to ferment our food. We are big believers that microbes are our friends, and most strains of E. coli are not only harmless but helpful to us — at least in our gastrointestinal tracts. But this strain, which is happy in the guts of many animals, is a real problem for humans. While ethanol levels above 3 percent kill it, the best way to avoid it is to use apples from the trees only and not from "the floor," where cows, sheep, or sneaky deer might have passed by. In fact, in the United States, the recent Food Safety Modernization Act Produce Safety Rule forbids using fruit that has unintentionally

come in contact with the ground from being used in cider unless significant steps are taken to reduce the pathogens, steps which are out of the realm of most home cidermakers. They include pasteurization, UV radiation, pulsed light treatment, ozone and high-pressure processing, among other treatments.

For this reason, we always recommend you pick ripe fruit directly from the tree. However, if that isn't possible and you want to shake the tree, place clean tarps underneath it and gather only the fruit that falls on the tarps (the pathogens come from contact with the ground). You will need to press soon after as there will be more bruising.

And of course, if you can't pick your own, you can buy boxes of apples. See the apple varieties reference in appendix 2 on page 308 for a listing of typical apples available and their characteristics to help you decide your blend.

SWEATING APPLES

Sweating is a process in which you store apples as cool as possible without nearing freezing for days, or even weeks, until they get a bit mushy. If the temperature of the apples drops below freezing, you enter the world of ice cider (see page 260) but we'll talk about that later. If it's near freezing, it's preserving the apple (think cold storage for all those apples you buy in spring at the grocery store that were picked months ago), which isn't what we want either. As the temperature goes up, the ripening process accelerates, and that's what we are looking for — a controlled continuation of the ripening process.

Ideally, you should store the apples in cardboard boxes or apple crates made of wood or plastic in a cool, dark environment. That might be a basement, a root cellar if you are lucky enough to have one of those, or we have heard of people buying used chest freezers and turning them up to just above freezing, then filling them with apples.

You may be thinking that putting fresh, crispy apples through the grinder makes sense, but sweating, not so much. But once you know why cidermakers do it, it makes a lot of sense. Remember when we told you that the starches in apples continue to be converted to sugars after they are picked? Sugars are good when you are making alcohol; we want our apples (or pears) to have fully developed sugars. By sweating them, we allow them to lose a bit of moisture, effectively

If you press your thumb into a sweated apple and it leaves a depression, the apple is done sweating.

condensing the sugars that remain. That's a good thing. In the process the apples get mushy, which will help them give up their juice in the press — also a good thing.

Time your grinding and pressing party for when you think the apples are going to be ready. How do you know if they're ready, you ask? An easy way is to grasp one of the apples in your dominant hand and press your thumb into it. If your thumb sinks in and leaves a depression, the apples are ready for the press.

That said, not every apple is improved by sweating, and different apple varieties have different optimal sweating times. In general, if the apple is a thin-skinned or an early-ripening variety, then it probably shouldn't spend a lot of time sweating. If you are someone who likes to dial things in, read on; otherwise just stick to the thumb test and it will all be good. Bruised and/or damaged apples already have the enzymes working on breaking them down and shouldn't be sweated, as you will likely get mold growth. If you just picked your apples and some have been bruised in the process, that's fine, just move them on to the next step the same day if possible. If you aren't getting to pressing for a few days, we recommend culling out all the bruised fruit for something tasty — apple pie, applesauce, apple butter — just not your juice.

As a rule, cider apples enjoy a longer sweat than culinary or sweet apples. Cider apples, like Kingston Black or Dabinett, can sweat for 3 to 4 weeks. Keeper apples, like Granny Smith, Northern Spy, or Baldwin, are improved by a couple of weeks of sweating, which allows all of their starch to finish converting to sugar. Dessert apples (those sweet little darling summer apples) don't do as well with sweating, and you should check in on them after a few days.

GRINDING APPLES

Also known as milling, the point of this step is to create pomace or mash that is ready for the press. There are two steps: washing and grinding. *Wait, you say, why are we washing apples that we freshly picked from the tree?* Rinsing them in a big sink or trough of fresh water removes any dust, debris, or leaves that might accompany them. Besides, a sink full of apples is a nice working environment for a couple of people, and it's also an easy way to look for and cut off small bad spots and reject any apple that you wouldn't eat fresh. If you are planning to make a wild yeast cider, don't worry — you won't wash off the natural yeasts, at least not enough of them to make a difference.[4] This is probably due to the fact that the apple's yeast floras are more than skin deep — they are in the bases, in the stems, and in the cores.

Do you need to cut and quarter the apples? It depends on the grinding/milling/smashing method or device you are deploying next. If the device can swallow whole apples, then don't worry about cutting them. If you are feeding your apples into a small juicer, you will want to cut them up to a size that will fit.

One of the first big purchases we made after moving to our homestead was a Happy Valley Ranch American Harvester two-tub cider press, which continues to serve our community. It's a lovely grinder and presser in a single, American-made device of wood and metal. Yes,

there are more efficient ways to grind apples than feeding them by the bucket into a hopper while hand-cranking a big flywheel round and round, as the machine grinds those apples down into the waiting wooden basket. A completely stainless steel (no wood to maintain) one-hopper, home-sized system with a grinder (each can be purchased separately) is also made in the United States and sold by Lehman's Hardware. After a while, you may see yourself transforming into something like a crab, with your cranking arm growing while the other shrinks away. But it's a piece of functional beauty that requires no power, other than human, and it naturally attracts people to it when it's in use. We think it's because it demands

human interaction — someone to feed baskets of apples into the hopper, someone spinning the handle around and around to grind those apples to pomace, and another person cranking down the screw press. Or, it is simply because in our high-tech world we are fascinated with something that harkens to a simpler time. Handwork aside, these grinders can be electrified, and there are many other systems that come power operated (see Resources for other equipment sources).

Our advice is to have your grinding serve your pressing. In other words, getting the juice is the main goal here, so you want the size and consistency of the pomace that your grinder generates to be a good fit for the press.

STONE FRUIT ARE THE PITS

We love adding other pome fruits (those with a core of small seeds surrounded by a tough membrane) we have harvested to our apples during this grinding phase to produce something special. You might think any fruit would do, but some are too soft (like persimmons) and don't juice well, as the pulp gets stuck in the bag. Stone fruit like peaches, plums, cherries, and apricots make excellent partners to apples in a cider, but their pits are, well, the pits. That's because when the pits are cracked, like when you run them through your grinder, they release a compound called amygdalin.

Amygdalin is a complex of cyanide, benzaldehyde, and glucose and can be found in the plant families of Caprifoliaceae, Mimosaceae, Oleaceae, and Rosaceae. Benzaldehyde is nice — it's also known as bitter-almond oil — and imparts pleasing almond aromas to the cider. Unfortunately, when we consume amygdalin it releases cyanide and, if in significant quantities of 0.5 to 3.5 mg per kg of body weight, can lead to cyanide poisoning. Amygdalin levels can be found in nearly any products made from theses fruits and seeds, including almond milk, nut butters, canned peach slices, and peach smoothies.

Apple seeds seem designed to pass through anything unscathed, and you can notice this when you grind and press your own juice. It is usually hard to spot more than a few nicked apple seeds in a pile of pomace. In studies, fresh-pressed apple juice has shown the highest levels of amygdalin, followed by juice that has been heat-treated, and

finally cider, which had no detectable levels of amygdalin after fermentation.[5]

While it's tempting to just chuck your whole stone fruit in with the whole apples in the hopper, don't do it, even though it appears the fermentation process degrades things like amygdalin. It's better to err on the cautious side. Take the time in the washing step to slice your stone fruits in half and quickly ditch the pit, or plant it. The world needs all the trees it can get.

Here's a pome pomace hack: Pears and other pomes featured on pages 146–63 have a softer texture, which makes it harder to extract the juices when pressing. To break up this texture and get the juice to flow, mix in a few cups of organic rice hulls (found at gardening supply stores) before pressing.

PRESSING POMACE

There are a few ways to look at pressing. At the farm or noncommercial level, it's all about batches. Grind up the apples to pomace, then press that pomace to sweet cider. In some cases, the mill and press are one unit, like with our American Harvester, so you grind a basket of pomace that is then slid a foot on the tray to be pressed for juice. Presses usually follow gravity, meaning they press horizontally downward, but bladder presses and some of the large-scale hydraulic rack and cloth presses work by pressing the pomace vertically.

Companies like Maximizer have separate grinders and presses, which trade out the wood slatted baskets of the American Harvester for stainless steel. Other companies like Harvest Bounty, Country Estate, and Glenwood have their own, slightly different stainless steel models as well. As you scale up you may move to continuous pressing, either on horizontal belts or on screw presses where the pomace doesn't wait around and is injected into porous cylinders, forcing the juice outward in a continuous motion. But pressing doesn't have to be about pressure. There are vibrating screens or centrifuges, although we have never seen these in operation.

Anyway, don't get lost in the technology — the point is to free the juice from the pomace with whatever you have available to you. A press is a wonderful piece of equipment to purchase and share in the community. If you are the handy type (or know someone who is), you might consider building your own. Another option is to approach your local cidermaker and ask if you can run a batch through their equipment at the end of their own pressing day or during a downtime.

1. **Wash.** Dump sweated apples into water and stir with your hands to separate leaves and twigs if taken directly from the orchard. Any apples that don't float should be composted or discarded.

2. **Prep the equipment.** Clean, assemble, and otherwise prep the grinder, press, and juice containers. Once the flow gets going it pays to have containers with lids washed and ready.

3. **Drop apples into the grinder.** Both manual and electric grinders have an optimal rate they like to be fed. On the electric model, start by slowly adding apples.

4. **Check the pomace.** As a rule, the finer the grind the more juice produced at pressing, as long as it's not too fine for the press, which would pass bits of pomace through to the juice.

5. **Press.** Depending upon the press, plan on applying pressure to the pomace for 15 to 30 minutes per batch. When juice no longer flows, dump the tub into a compost bucket and rinse the bag for the next batch from the grinder.

6. **Catch the juice.** Depending upon the apples you are using, expect between 2 to 3 gallons of juice per 40-pound box of apples.

The next time we had mounds of raw peelings and cores from making dried apple rings, she made our first fruit scrap vinegar. It was okay, but not great. A few years later, when we understood more about vinegar's previous life as alcohol, it improved. The acetic bacteria feed on ethanol, not apple juice. That means you need alcohol to make vinegar. The cores and peels themselves don't have enough sugar and nutrients to feed the yeast, so instead of adding yeast (some old-school recipes add bread yeast to begin the process), you will add sugar and let the wild yeasts that are already on the apple skins go to work. To that same end, when we started to make hard cider, we realized that the pomace made an excellent vinegar.

Pomace vinegar is a wonderful way to upcycle at least some of cider's by-product. As far as adding nutrients, we found it depends on the fruit and the amount of sugar it releases into the soaking water. Because in most cases there is still plenty of fruit in the pomace, unlike just cores and peelings, you may not need to add a sugar source. For example, pear pomace makes our favorite vinegar; the fruit retains enough sugar after pressing and doesn't need any added sugar, and the pears also retain the unfermentable sorbitol so it will stay lightly sweet. We splash this in sparkling water for a refreshing drink. Sweet apples, like Golden Delicious, generally have enough sugar for the pomace. We include sugar in this recipe to take the guesswork out of the process — the added sugar will guarantee there is enough ethanol to get a nice acidic vinegar. The percent acidity of the vinegar is proportionate to the ABV. An ABV of around 5% will give you a vinegar that is around 5 percent acid.

POMACE TO VINEGAR!

At the beginning of our homesteading adventure, Kirsten preserved anything that wasn't nailed down. It's no wonder that when she read *Farmer Boy* to our children and came across a line about apple core vinegar, she was intrigued. Armed with the Foxfire books and Carla Emery's *The Encyclopedia of Country Living* (but not the Internet), she got closer to figuring out how to make this happen.

Pomace Vinegar

This is a simultaneous fermentation, which means the yeasts will turn the sugar into ethanol, and the acetobacters (from the addition of raw vinegar) will be waiting in the wings to get started. We add hot water to help draw out the juice, much like steeping tea. Despite the hot temperatures, we have never found the wild yeast development to be a problem. Use a lower temperature water if this is a concern to you.

As for the sugar, the yeasts don't care what you feed them — white sugar or molasses or honey — but molasses and honey will impart some flavors in your final vinegar, which can be good or bad depending on your goal. You can use this recipe for many kinds of fruit scraps, or mashed overripe fruit. It is helpful to have a thermometer strip that sticks to your vessel — like the kind that goes on an aquarium — as well as a seed-starting mat if you'll be making this in cooler temperatures.

**YIELD:
1½ QUARTS**

- 1 pound (450 g) apple mash from pressing
- ¾ cup (155 g) sugar, any kind
- 2 quarts (about 2 L) boiled water, cooled to 160°F/71°C (unchlorinated if possible)
- ½ cup (118 mL) raw unpasteurized unfiltered cider vinegar or a vinegar mother

1. Combine the fruit and sugar in a ½-gallon jar. Pour in at least 1 quart of the water and continue to add more until it reaches the neck of the jar. Let cool to room temperature, add the raw vinegar, and stir well with a long-handled spoon.

2. Cover the jar with an oxygen-permeable cloth or paper, such as a coffee filter, butter muslin, a tight-weave cheesecloth, or a tea towel. Secure with a string or a rubber band or screw the ring from the jar over the cloth to keep out fruit flies. Place the jar in a spot that is around 75°F/25°C to 86°F/30°C (82°F/28°C is its sweet spot).

3. Week 1: Stir once a day for the first 5 or 6 days. You may see bubbles; that is good. Unlike most ferments, you want to get some oxygen in the mix. Stirring will also help manage the fruit scraps that poke out of the liquid, as it will resubmerge and suffocate any yeast growth while giving the acetic bacteria a place to breathe.

4. Week 2: Continue to stir occasionally with a wooden spoon, if you remember (it is not a big deal if you don't). You may see some yeast growth but it is harmless — stirring will help combat that. The ferment will slow down around the end of this second week. Remove the cover from the jar — you may see a transparent film developing on top. That is the beginning of the mother (mother of vinegar or MOV). Remove it, or pieces of it, and set it aside. If there is a white or milky film of surface yeast, do not save. Strain the contents of the jar through a sieve into a clean jar. Add the mother to the clean jar with the liquid and cover the jar. Discard or compost the solids. Let the jar sit for another 4 weeks.

continued on next page

Pomace Vinegar

continued

5. Week 6: Taste the vinegar. At this point it should have nice acidity, but the vinegar will take another 4 to 8 weeks to be fully developed. The timing is largely influenced by oxygen and temperature. If it is cooler than the ideal range, it will take longer.

6. Weeks 10–16: Bottle the vinegar, saving the mother for another batch or sharing her with a friend. It is important to bottle the vinegar to remove the exchange with oxygen because if left too long, the vinegar can eventually weaken (become watery) as the breathing bacteria continue to eat.

7. You can use it immediately, but honestly, letting it age even more will mellow it out and give it a smoother flavor. When bottled or barreled without oxygen, vinegar can be aged indefinitely. This vinegar cannot be used for canning because without testing you cannot be assured that the percent acidity (see note) is 5 percent — the minimum that is safe for canning.

NOTE: Once the fermentation is finished, there are two ways to measure the acidity in vinegar: percent acid and pH value. Percent acid is the number of grams of acetic acid per 100 mL of vinegar — in other words, 100 mL of 5% vinegar has 5 grams of acetic acid. It is measured using a titration process. The pH scale measures the strength of the acid and can be tested using simple pH strips or pH meters. They both have a place in measuring but are not interchangeable. This means that there is not a pH measurement that corresponds with a percent acid. Different vinegars can have varying pH values based on the base fruit, even though they may have the same percent acidity. Percent acid is predictive by the ABV. For example, 5 percent ABV will be around 5 percent acid. For this, be sure to take a specific gravity reading before fermenting. You don't need to know either measurement to make a safe vinegar for flavor and enjoyment.

BLENDING JUICES

Blending is where the magic starts. Until this point, cidermaking has been a pretty physical endeavor: pick, sweat, squash, squeeze. Now we are going to use a lighter touch and possibly make a few changes that will pay off months down the road, when we are opening and enjoying these ciders.

Why blend? Well, as we learned earlier, very few apples have the right mix of sweetness, tartness, and tannins — from a flavor perspective — as well as the qualities needed to produce a cider with the right alcohol and pH level. A blend will give you favorable amounts of acid, sugar, tannins (or phenolics), and aroma compounds. If you are planning to age your cider for a few months to several years, kicking up the tannic apples can provide preservative properties. The trade-off is that unless they are bittersharps, their pH is higher than our target (ideally 3.3 to 3.6, with a max of 3.8). By measuring the pH at this step, you will know if your cider is in the target window. If it's not, you will want to blend in other varieties to bring the pH within the window, if you can. You may also need to add some sugar if your apples don't have enough sugars to reach the level of alcohol desired. That desired level could either be the alcohol level you want in your finished cider for drinking pleasure, or a minimum level to preserve the cider in the bottle. Generally, that level is 6 percent ABV, which translates to a specific gravity of 1.045.

Three different freshly pressed apple ciders before blending

GOLDEN RUSSET
SG 1.075 pH 3.2

GOLDEN DELICIOUS
SG 1.070 pH 3.9

LIBERTY
SG 1.059 pH 3.5

Low Specific Gravity

What if the specific gravity of your juice is less than 1.045, the minimum needed to produce the ABV that will preserve your cider naturally? First off, know that if you are measuring with a hydrometer, they are often calibrated to solutions at 60°F/16°C, so unless your cider is at that temperature, it will be off a bit (for instructions for measuring specific gravity, see page 73). Secondly, the gas bubbles will attach to the bottom and sides of the hydrometer, giving it a boost upward, artificially raising the specific gravity reading. We shake or swirl the hydrometer in the cider to dislodge the bubbles from the bottom and sides. This also helps knock down the bubbles on the surface of the cider, to get a more accurate reading. If you have adjusted for these things and are still getting a reading below 1.045, you need a sugar source to bump up your reading.

JUICE BLENDING

Blending in another juice with a higher specific gravity (more sugar) at this step is ideal and a good reason to make at least a small pressing run of your sweetest apples, so that you will have something on hand to give your cider a sugar shot.

CHAPTALIZATION

You will need to add sugar until you get your raw juice to the target specific gravity. This process is called chaptalization and it's different from back-sweetening, which happens at the end of fermentation. The trick here is to use a small amount of your juice to determine the necessary ratio of added sugar to juice, then scale up that number to the amount of juice you are working with. On the next page is the process for a 1- or 3-gallon batch, but you can easily alter it to the amount you are making.

One last word on chaptalization: this process is far from precise. Going through the math on page 71, you would think that you have it dialed in, but hydrometers aren't exact and are calibrated for an exact temperature of the liquid. It can also feel like an eye test when you try to precisely match the level of the cider to the lines on the hydrometer. Plus, the sugar actually throws things off a bit and your final specific gravity reading might be a higher estimate than what you get in your finished cider, but it will be close — in most cases, it will probably be off by less than 0.5 percent ABV.

High Specific Gravity

What if the specific gravity of your juice is too high? Some of you, like Christopher, might be wondering how this could be a problem, but if your target style has an easy-drinking alcohol level — say around the minimal 6 percent — then it's quite possible your juice is going to overshoot this alcohol level by as much as a few percentage points. To bring the ABV back down to your target level, you can either stop the fermentation before it reaches full dryness, or you can let it go until fully fermented and bring the alcohol level down by diluting with fresh water, a process known as amelioration. Stopping the fermentation prematurely at the right point requires a little bit of math, taking good notes, and remembering to

HOW TO SWEETEN YOUR JUICE

1. Separate out 1 quart of juice and bring it to 60°F/16°C.

2. Add some of the juice to a hydrometer jar and float the hydrometer to take a reading. Note both the specific gravity reading and the likely ABV when finished.

3. It's time to do some math! General guidelines are for every 0.005 increase in specific gravity, which is about a 0.64 percent increase in ABV, you should add 9 teaspoons of sugar to every gallon of juice. Since you are working with a quart (¼ of a gallon), that works out to adding 2¼ teaspoons of sugar. So, for example, if our specific gravity reading was 1.040 and we want to raise that to 1.050, you would add 4½ teaspoons of sugar to your test quart of juice.

4. Pour the juice from the hydrometer jar back into the quart jar, add the amount of sugar you calculated for your situation, and stir to combine.

5. Add some of the sugared juice to a hydrometer jar and float the hydrometer in it again.

6. Note the rise in specific gravity and repeat the process if necessary until it reaches your target. At this point you know how much sugar you added to your quart of juice, so it's time to figure out how much sugar to add to your 3-gallon carboy or 1-gallon glass jug. To do this, take the amount you came up with for your quart and multiply that by 3 for a 1-gallon jug or 11 for a 3-gallon carboy, since you have already added sugar to 1 quart. In our example, if you

added 4½ teaspoons to your quart, then if you are using a gallon you would add 3 times that, or 13½ teaspoons (⅓ cup), and if you are working with a 3-gallon carboy it would be 11 times that, or 49.5 teaspoons (about 1 cup).

7. Add the test juice back to your carboy or glass jug to reunite it with its new, sweeter version.

SPECIFIC GRAVITY, BRIX, AND PREDICTING THE FUTURE?

This is a good point at which to talk about specific gravity and Brix, which both measure the sugars present in juice and cider. Specific gravity describes the relative density of the juice compared to water. Water has a specific gravity of 1.000, so any sugar present raises that level. Apple, pear, and other juices have specific gravities that usually range from 1.045 to 1.065.

In the world of wine, it's about Brix, which measures sugars in grapes. But apple juice, being about 90 percent water and 10 percent soluble liquids, doesn't do so well with Brix refactors, which are calibrated against pure sucrose. Once fermentation begins and there is the presence of ethanol, Brix refactors don't work anyway. Luckily, simple and inexpensive devices like hydrometers give you both readings, and they work really well for cidermakers. If you are curious and want to compare notes with a winemaking friend, there are tables to convert specific gravity to degrees Brix; a specific gravity of 1.040 equals 10.0° Brix.

Why is it important to know the density of your juice? Because as the juice ferments, the sugars are converted to ethanol and carbon dioxide in roughly equal amounts, lowering the density of the juice. By measuring the specific gravity during the fermentation stage, you can get a pretty good idea of how much of the sugar has been devoured by the yeasts and converted to alcohol. When your reading is at 1.000 or below, you can feel pretty confident that all the work has been done. You might be thinking, "How could it get below 1.000, which is the baseline of water?" It's because ethanol has a lower specific gravity than water, so on higher alcohol ciders that are fully fermented to dry, you will get readings below 1.000.

check the specific gravity regularly to catch it at the right point. The easiest way is to get out of the way of the microbes and let them do their thing to full dryness and then dilute the cider with water before bottling. To do this, use your initial and final specific gravity readings to determine what your final ABV is likely to be, then add enough water to reduce the ABV to your target level. For example, let's say that your initial specific gravity measurement was 1.070 with a potential ABV of nearly 9 percent. If you want to knock that back to about 8 percent, which is in the range of the stronger commercial ciders but not in the range of apple wine, then divide 8 by 9. This gives you the percentage of juice you want in the cider (89 percent), so you want to add 11 percent water. For a 3-gallon carboy, you would add about ⅓ gallon of water to bring the ABV down from 9 percent to 8 percent. Note that you probably won't notice a difference in taste of your cider up to 15 percent water.

1. **Assemble everything.** You will need a wine thief to extract the cider for testing, a hydrometer jar, and a hydrometer.

2. **Add cider, then the hydrometer.** Pour cider into the hydrometer jar, then measure the temperature before carefully dropping the hydrometer into the cider until it floats. The cider should be close to 60°F/16°C for the most accurate reading.

3. **Measure accurately.** For an accurate measurement, the hydrometer needs to be free-floating and not stuck to the sides of the jar. If it is stuck to the side, give the hydrometer jar a gentle spin to free the hydrometer. Read the measurement level at the bottom of the curvature in the cider, called the meniscus.

While it's best to get your apples or raw juice from a local grower, that isn't always realistic. Fortunately, you can make good cider with just about any juice.

Apple juice found in the refrigerated section of the grocery store can be one of the best choices. These juices are as close to fresh-pressed as you can get at a grocery store, and they are still more affordable than buying a case of apples at retail to press. They are pasteurized (labeled flash-pasteurized, just pasteurized, or HTST) but usually they are not ultra-pasteurized (labeled UT, UHT, or UP), which means that the flavor will be more robust. You have the greatest chance of getting some acidic and bitter notes, not just sweetness, in fall, when you often can find juices made more regionally. Read the label carefully, though, because even though the store is going to the expense of refrigerating the juice, it doesn't necessarily mean that it needs to be. If the juice has preservatives like potassium sorbate or sodium benzoate, it's being refrigerated for looks; pass it by for your cider because those preservatives that have kept microbes out will likely continue protecting the juice from the yeasts you intend to introduce.

Apple juice found in the grocery store aisles is made to be shelf-stable for years. This means it was ultra-pasteurized (exposed to very high temperatures), which devastates all microbes and likely any subtle aromas that might have been there. These juices are sweet and sometimes even taste a bit cooked. When we combed the aisles, we found very few juices that actually contained preservatives beyond ascorbic acid (vitamin c). We don't worry about ascorbic acid, but we do avoid potassium sorbate and sodium benzoate as mentioned above. This is our least favorite on this list, but we have made some respectable ciders in a pinch with these juices. For more noteworthy flavors with these juices, you may choose to infuse them, which you will learn about beginning on page 223.

Organic apple juice in glass bottles is often a little cloudy, and thicker toward the bottom. It clears nicely with fermentation. It is significantly more expensive than nonorganic apple juice, though it comes in its own fermentation glass jug which you can top with an airlock cap and use it immediately (and reuse). Check the label for preservatives and, as mentioned above, avoid the ones that have preservatives that would interfere with fermentation microbes in your cider process.

Clear 100% real juice from concentrate, 100% cider from concentrate, 100% apple juice cider, organic from concentrate are all names for the classic, crystal-clear, light amber juice we remember from school lunches. It is the most common type of apple juice sold in grocery stores and comes in many brands. These juices make a surprisingly decent cider. They can become more interesting with the addition of wild yeasts or other juices. They are generally sold in plastic bottles and are often the most economical option. Again, check preservatives.

Natural, unfiltered, cloudy, 100% pure unfiltered, 100% pure pressed are all terms that were the most confusing to us. These "natural" words were followed by *apple juice* or *apple cider* (which incidentally means nothing about what is inside). Some of these were truly just bottled juice with nothing added. These are good choices that clear nicely during fermentation and can make respectable ciders.

There is one type of juice to be wary of. We bought 20 gallons of juice in the medium price range that was labelled "pressed from fresh whole fuji and gala apples . . . 100% pure unfiltered apple cider not from concentrate." Sounded good. The hint should have been the opaque, even-toned butterscotch color. We seriously couldn't tell if the color was coming from the bottle or the juice. *You don't want this.* (Look, instead, for uneven sedimentation and colors.) Apple juice that is too processed (put through a giant centrifuge and a machine that removes some of the color to keep it just so, among other things) will have solids suspended in it, and it will not clarify *at all* during fermentation. The flavor yields cider that is subpar to undrinkable, in our opinion.

Apples in orchard with yeasts visible on skins

PITCHING YEAST

Yeasts are going to do the heavy lifting in the upcoming fermentation step, so this is the time to pick your team. If you are going to go natural and depend upon the wild yeasts on your apples or blossoms, then you are already set. If you are going with yeasts other than the locals, then you need to do two things. First, you need to decide if the yeasts you are introducing (called pitching) can outcompete the locals. In our experience, commercial yeasts have no problem taking over the fermentation from the wild yeasts, so we almost never need to add sulfites. If you are concerned about the microbes on your apples, the first step is to take out all the locals by introducing sulfites, in the form of sulfur dioxide (SO_2).

Selecting a Commercial Yeast

What yeast is the best one for the cider you are about to make? The answer is: either your go-to favorite, or "it depends."

GO-TO FAVORITE

We know a lot of good cidermakers, both amateur and professional, who consistently use only one or two strains of yeast in every cider they make. Why? Because, they will tell you, it works. Given all the variables in cidermaking — source of juice, sugar, acidity and nutrient levels in the juice, fermentation temperature, to name a few — they have found a strain that consistently works for them. We are the same way, as you will read next.

FINE-TUNING YOUR YEAST

You can fine-tune your yeast selection based upon a number of variables. Our suggestion is to pick a couple of elements that are most important to you, then try out a yeast that is designed to excel at those elements. By way of example, here are the top four things we optimize around and the yeasts that we use consistently because of their performance:

- **Low nitrogen needs.** We make a lot of wild ciders, which we have found don't need additional nitrogen. To keep things simple, we remove the need to measure yeast assimilable nitrogen (see box on page 88) by looking for yeasts with minimum nitrogen needs.

- **Low temperature range.** We have only one cider cave to ferment in and we do a lot of wild ciders, which like it on the cooler side

(50°F/10°C to 65°F/18°C), so we look for yeasts with a lower temperature tolerance in this range.

- **Malolactic fermentation (MLF) compatibility.** We generally want MLF to take place, so we look for yeast strains that play nice with the lactic acid bacteria. Specifically, we look for those that have a low nutrient demand, produce low amounts of sulfur dioxide, and autolyze well so that their little dead yeast bodies become nutrients for the bacteria.

- **Strong fruit aromas.** Some yeasts are strong and have other characteristics that many cideries look for, like low production of volatile acidity, hydrogen sulfide, and sulfur dioxide, and low foaming. But these yeasts sometimes step all over the fruit notes — we are looking at you, EC-1118, also known as Prise de Mousse (and one of the few you will find in your local brewing supply shop) — so we choose other yeasts to keep the fruit aromas in our cider.

Based upon these four criteria, our go-to yeast strains are Lalvin ICV D47 and 71B.

To Sulfite or Not to Sulfite

Sulfites date back to at least the Romans, and there are pros and cons to their use. Why use sulfites? In a word: predictability. By wiping out most of the natural yeasts and bacteria, you reduce the competition to the commercial yeasts that you want to thrive. Wild yeasts (see box on page 81) can make some great-tasting ciders (more about this on page 131), but you don't always know what you're getting — at least the first time. And some of the bacteria present on the apples are not always helpful for making a great-tasting cider. Bacteria are more susceptible to the effects of sulfur dioxide in small doses than some yeasts, so at low doses it's possible to target all the bacteria — including acetic acid bacteria, lactic acid bacteria, and *Zymomonas* species — and some of the yeasts. In our experience, it seems that many

CLEARING THINGS UP WITH PECTIC ENZYMES

Fruits like apples are made up of cells that have structure — just like tiny rooms in a big building. Those cells have walls and those walls are made up of stiff things, like cellulose, and binding things, like pectin. When you grind up the apples, you reduce those neatly organized rooms to rubble, which lets you draw off the juice when you press. Though most of the cellulose remains in the pomace, the pectin follows the juice, which is usually fine. Sometimes, though, there is too much of a good thing and you might want to clear some of that pectin out of your juice. To do that, you need some enzymes that love to break down these pectins. You can buy small bags of pectic enzyme powder in brewing supply shops or online. You don't need a lot — between ½ and ¾ teaspoon per gallon per the manufacturer's instructions.

of the commercial yeasts, including the ones that say they might have difficulty competing with wild yeasts, do just fine — even when pitched several days after their wild cousins have had full access to the juice. So even if you will be pitching yeast, this addition of sulfur is not necessary.

The are four top reasons to avoid sulfites. The first is that lactic acid bacteria will be harmed by sulfur dioxide, which can be limiting later in the process when you might be counting on them to generate a malolactic fermentation that will mellow out the acids in your cider. The second con is that many people are sensitive to sulfites and the fermentation process already typically produces a small amount of sulfur dioxide, depending upon the yeast strains involved. Thirdly, if you are aiming for a live, probiotic drink, you can't get there if you kill off all those natural microbes in the beginning. Finally, sulfur dioxide will also prevent proper fermentation if you choose to make vinegar down the road.

Controlling Alcohol Content (ABV)

During fermentation, the yeast metabolizes the sugar in the must (cidermaking term for the juice you will be fermenting) to form equal amounts of carbon dioxide and alcohol. Measuring the sugar content of the must through its specific gravity gives us the potential final alcohol by volume (ABV) if the yeasts metabolize all the sugars. Whether they do or not depends on three things: the alcohol tolerance of the yeasts, the starting amount of sugar in the must, and if we,

as cidermakers, intervene in the natural process or not.

As mentioned earlier, we recommend a minimum ABV of 6 percent to assure that your cider has enough alcohol to preserve it until you are ready to drink it. Most apples naturally want to ferment to between 6 and 8 percent ABV, so this is a natural sweet spot. If you'd like to boost the ABV beyond what the natural sugars in must are capable of, the limiting factor will be your yeasts. They will party hard, but like us they all have their limits, and at some point they can't take the level of increasing alcohol and they die. Some, like many wild yeasts, knock off early in the 2 to 5 percent ABV ranges. Others — you might be thinking of human manifestations of these guys — keep partying long after everyone else is under the table, but even they have their limits.

The issue with wild yeasts is they are like amateur drinkers that just heard about the party and stopped by. They're not reliable but have a lot of fun. As always, it depends on your cider goals, or to keep with the analogy — the kind of party you are looking to host. In general, the *Saccharomyces* clan stay at the drinking table much longer than their wilder cousins,[6] and they are professionals. So much so that they are bred in laboratories around the world to do this one thing and they do it very well. If you have an ideal ABV in mind, you need to make sure there is enough sugar and that your yeasts can take you there, which usually means at some point depending upon *Saccharomyces* strains that are either purposely introduced or get to the party uninvited.

1. This step is optional and is only for those who choose to start with sulfur dioxide (we do not to do this.) Follow the manufacturer's instructions to add sulfur dioxide in a dose of 50 parts per million to the juice. This will need to sit for 24 hours before pitching the yeast.

2. Add the appropriate amount of active dry yeast to unchlorinated water (see box on page 80) heated to 104°F/40°C.

3. Stir the yeast into the warm water until fully incorporated and wait for 30 minutes.

4. Pour the hydrated yeast mixture into the apple must.

WATER MATTERS

When rehydrating commercial yeast, there are three things about the water that are important:

1. **MINERALS.** The water needs to have natural minerals because if there are fewer minerals in the water than in the yeast, the water flows toward the minerals in the yeast, bursting the yeast bodies, which kills the yeast that your cider is depending upon for fermentation. That's why you should not use distilled water or other type of water that has had the minerals removed.

2. **SAFETY.** Antimicrobial agents like sulfur dioxide and chlorine do a good job of killing things like yeast. You used to be able to reliably boil or leave out overnight a pan of chlorinated city tap water to produce "clean" water, but with the use of chloramine, this no longer works. If your tap water is chlorinated, it's possible the chlorine is at a low enough level not to affect the yeast, but you can only know for sure through trial and error. It's better to buy good-quality spring water with natural minerals to use for your rehydration.

3. **TEMPERATURE.** Temperatures around 104°F/40°C are ideal for breaking down the hard, outer casing of the dry yeast while not damaging the cell within. Recipes that call for pitching yeast directly on the must will likely result in high yeast die-off, which often still works — but why reduce your numbers if you don't have to.

Ideally the sugar, fermented to dry, will land exactly at the level of alcohol you want. If so, your job as cidermaker is to stay out of the way and let the yeasts do their thing until they are done. If the sugars are too high and your yeasts can go higher than the ABV you want, then you will need to chaperone the party, and at the right point (when your target ABV has been reached) you will need to send everyone home, which in cidermaking means either expunging your yeasts or killing them off.

LESS THAN 6 PERCENT ABV

The easiest way to stay under 6 percent ABV is to start with a must with a specific gravity of less than 1.046 or a Brix of 11 or lower. Since sugar develops in the apple as it matures, you can have a fully mature variety at a Brix of 10 or an immature apple caught on its way to a much higher Brix at maturity. You want the former. The majority of apples are going to have a Brix above 10 so you might need to blend in a lower-sugar fruit, like guava (7.7), strawberry (8.0), or gooseberry (8.3),

MEET SOME OF THE WILD YEASTS

Apples come with a wide range of yeasts, most of which don't directly contribute to the fermentation process. Some of the yeasts you will find specifically on apples are *Kloeckera apiculata*, *Torulopsis stella*, *Saccharomycodes ludwigii*, and *Candida* species. What's missing? The main fermenting yeast genus, *Saccharomyces*. It is rarely found naturally on apples but is likely already waiting on your cider equipment and in your cider room if you have made batches of cider with commercial yeasts in the past. Yeast like *Saccharomyces* can exist for months in cracks and crevices of clean, dry equipment, like the wood in a press or scratches in plastic buckets.

In wild yeast fermentation, it's usually the *Kloeckera apiculata* team that moves first and grows rapidly, but it doesn't last long; it dies out after the ethanol reaches about 2 percent. That's when the *Saccharomyces* teams like to take over.

or you will need to dilute your apple juice to bring down the sugar levels. A third option is to rely on yeasts that are alcohol lightweights, then you will likely need to starve them. Let us explain.

Nearly all commercial yeasts are going to have an alcohol tolerance above 6 percent, so if you allow them to do their thing naturally you will get higher alcohols than you want. If you stop adding nutrients, you are at least not feeding them, which will start to slow them down. Still, you will need to monitor the specific gravity during primary and perhaps secondary fermentation, and at the point where you are getting consistent readings of 1.042 or less you will intervene to remove the yeasts. Sulfur dioxide and heat are the two easiest ways to do this, but both have ramifications for bubbles and sweetness since you will have no yeasts left for further fermentation to create bubbles and you will likely be leaving sugars on the table.

6 TO 8 PERCENT ABV

No special attention is needed here. In our experience, both purchased apples and unpreserved juice will produce cider within this range using nearly any commercial yeast.

ABOVE 8 PERCENT ABV

None of the wild yeasts that we are aware of play in this space — they will die off at this ethanol level. To make these high-ABV ciders, you'll have to turn to commercial yeasts. Most of the commercial dry yeasts, from companies like Lalvin, Red Star, Vintner's Harvest, or White Labs, have an alcohol tolerance threshold in the 12 to 16 percent range. Each of these labs have at least one yeast strain, and in many cases several, that can reach 18 percent (that's a must with an initial specific gravity in the 1.138 range!). At this level you will also need to develop a yeast nutrition feeding regimen because they will need nutritional support for all that activity.

Higher still? Since you asked, you can go higher with the help of some very special yeasts like White Labs WLP099 Super High Gravity Ale Yeast, which clocks in at a whopping 25 percent ABV. That's 50 proof. Think Campari or limoncello. At this point, you are fermenting more introduced sugar than fruit must, aren't you? The issue here, and the trick to mastering a very high percentage ABV cider, is managing for flavor. Specifically, that means keeping the herbal/citrus/fruit aromas of the must and providing enough time in the secondary fermentation and aging steps to allow for those secondary and tertiary flavors to fully develop and some of the sharp edges to smooth away.

FERMENTING

Finally, we are fermenting! This is when the yeasts convert the simple sugars in your sweet cider to alcohol (ethanol) and carbon dioxide.

GUIDELINES

Let's start with some basic guidelines:

Good hygiene is your friend. If you get in the habit of installing some good practices now, you can continue doing them in later steps and will have a better chance of success.

Air is not your cider's friend — at least not until you get past making a great hard cider and move on to vinegar.

Temperature matters and will affect your cider in predictable ways.

Yeast is your partner. You cannot make cider without them, but not all of them are helpful.

Let's look at each of these in a little more detail.

Above: Pouring fresh-pressed cider into carboy

Left: Foam and gross lees pushing up through carboy neck during primary fermentation

HYGIENE

There are two parts to hygiene: care and cleaning.

To keep things clean, we lean toward sanitizing, as sterilizing isn't realistic for most home cidermakers. Sterilizing is any process that removes absolutely all microorganisms. Sanitizing is cleaning and then treating in a way that gets most of the microbes, but some may remain. There are several products available to home brewers to sanitize equipment. At our scale, which is about 120 gallons or so per year of cider and maybe 12 gallons of perry, we are all about no-rinse solutions. Oxygen-based cleansers like One Step and acid-based sanitizers like Star San will clean everything we need to clean in our cidermaking operation. Stronger cleaning options using caustic detergents are available, and you can bump up your disinfecting game with peroxyacetic acid. The problem we have with both of these solutions is that they are hazardous chemicals, and all our floor and sink drains feed our gray water system in the orchard. We don't want to be around them while cleaning or out in our orchard. If your cider operation is home based, you should think about adding the no-rinse nontoxic oxygen and acid-based products. They are safe, and it's easy to mix up just the quantity that you need.

When we are making cider and need to clean an airlock, for example, or quickly sanitize our wine thief, we tend to spray it with a neutral vodka, which is widely available and identifiable. We also use straight vodka in the water reservoir in the airlock as an extra measure in case fruit flies or other spoilers land in this liquid, in the off (but possible) chance that a vacuum created in the carboy pulls some of this in.

You might be thinking the stronger the better and are tempted to get the really strong stuff, like Everclear and other 180-proof liquors. We thought the same thing until one summer when we were at a picnic with some local winemakers and Christopher brought up the subject of spraying things down with alcohol and asked how much Everclear they go through in a year. That's when we learned that very high alcohol doesn't work as well because it evaporates too quickly. You need enough "contact time" for the alcohol to do its job. We look for 100 proof (50% alcohol by volume) liquor that's affordable because, alas, none of it ends up consumed by us.

AIR

Air, and more specifically oxygen, is not your friend. In the very beginning, right after you've pitched the yeast and before fermentation gets going, air is not a problem — it's actually helpful to the yeasts during their initial explosive growth. You can simply cover your bucket with a clean cloth or, if using a carboy, stuff a wad of cotton in the mouth or place a piece of plastic wrap over the opening. After this brief phase, though, air has the potential to hold both unwanted microbes and oxygen for aerobic microbes already present in the juice that we do not want to see prosper. As a general guideline, you want to minimize the headspace in your fermentation containers and keep new air from entering this headspace, which will quickly become carbon dioxide and displace the ambient air once the fermentation gets going.

TEMPERATURE

Keep it cool, or at least on the cool end of your yeast's ideal temperature range. Temperature is one of the best tools we have to control the microbes' work during fermentation. Cider is no exception, as temperature has a direct effect on the flavor and aroma of your cider. Each microbe has a sweet spot in which it will thrive, so temperature influences which microbes will flourish and is a method to speed up or slow down yeast. For example, you will use cold temperatures (called cold crashing) to stress and eventually stop the yeast from consuming all the sugar in ice cider, leaving some for us to taste in the final product. Strive for fermentation temperatures in the mid-50s°F/10s°C to mid-60s, unless you are using a commercial yeast that requires a higher temperature range. There are some commercial yeasts that start in the mid-60s and go up into the 80s, so if that's the one you are using, adjust accordingly. If it's colder than 50°F/10°C, the yeast may become stuck, meaning it no longer continues to process the sugar into ethanol and carbon dioxide. At temperatures higher than its range, the yeast may be outcompeted by other microbes that work better at higher temperatures or it may become stressed, which can lead to an excess of sulfur being produced and some very disagreeable aromas.

MICROBES

First off, there is no cider without microbes — at least not yet, and we won't be lining up when the first lab-produced sterile ciders hit the store shelves. There are three schools of thought when it comes to microbes and cider. The first, exemplified by cidermaking couples Bill Bleasdale and Chava Richman of Welsh Mountain Cider (page 143) and Jonathan Carr and Nicole Blum of Carr's Ciderhouse (page 138), as well as Sean Kelly of WildCraft (page 230), is one of a very light hand. Orchards are no-spray and ciders are produced without sulfites or commercial yeasts. The second, exemplified by Peter Mitchell (page 290), employs a heavier hand in the fermentation process. After initially applying sulfites to reduce the native microbial load in the juice, specific commercial yeasts, chosen for their unique characteristics and dependability, carry out the fermentation. The third is somewhere in the middle, exemplified by Blair Smith of Apple Outlaw (page 186), who recognizes the benefits of both approaches and produces mostly controlled, commercial ciders, but also features crafted wild batches.

Fermentation Phases

There are four stages of fermentation: pre-fermentation (before the microbes really start working), primary fermentation (when the microbes really get going), secondary fermentation (the slow-burn phase, after all the fireworks have ended), and malolactic fermentation (bacteria lowering the total acid to produce a cider with a more complex flavor).

THE MICROBIAL COMPONENTS OF FERMENTATION

The fermentation of apple must as it becomes cider is a complex series of microbial reactions involving many genera and species of both yeast and bacteria. The ultimate flavor profile of your cider is determined by the type, ratio, and dynamic interplay of these microorganisms that you will never see. Yeasts are primarily responsible for the first alcoholic fermentation, and lactic acid bacteria lead to the second, malolactic fermentation phase.

YEASTS. It is often suggested that if you use commercial yeast, you should use sulfur dioxide to remove indigenous yeasts and bacteria so that the commercial yeast, chosen for desired flavor or process characteristics, can be assured little competition. These commercial yeasts are of the genus *Saccharomyces* and have varying but typically higher degrees of tolerance to sulfur dioxide, so they will be unaffected. Before you kill the natives, know that many cidermakers, including us, believe the indigenous yeasts contribute to the ultimate flavor profile of the juice.

A packet of commercial yeast can be comforting, because we can "see" and measure them, while wild yeasts are invisible. Commercial isolated yeast strains are often employed early in the apple must-to-cider process to assure a reliable outcome, but you don't need to use commercial yeast to make a successful cider.

BACTERIA. Lactic acid bacteria can remain pretty constant through the alcoholic fermentation while the yeasts are dramatically scaling up and then back down as they chase sugar. Acetic acid bacteria are another thing. Because they are aerobic bacteria, they need air to survive, so the idea is to keep air out. However, a slight dose of acetic acid can be part of a delicious sour cider or, conversely, be a spoiler turning the whole batch into unintended vinegar. How the apples are ground, the length of time until pressing, and the amount of time taken to fill the fermentation container all play a role in if and for how long acetic acid bacteria are present in the alcoholic fermentation process.

TEMPERATURE AND CIDER FERMENTATION

LESS THAN 50°F / 10°C: Slow fermentation, especially good for pitched yeasts.

LESS THAN 59°F / 15°C: Ideal upper limit for fermentation, especially if using native yeasts.

59°-68°F / 15°-20°C: Many of the commercial yeasts prefer this temperature range; it's a little warm for wild yeasts.

68°-77°F / 20°-25°C: Acetic acid increases, resulting in loss of fruity aromas and quality.

77°-86°F / 25°-30°C: Yeast's ability to convert sugar to alcohol is impaired.

86°-104°F / 30°-40°C: Cider flavor is very negatively affected.

GREATER THAN 104°F / 40°C: Yeasts die.

PREFERMENTATION (YEASTS CONSUMING OXYGEN)

You could think of this phase as the calm before the storm, the pregame warm-up, or the afternoon sound check before the evening live concert — you pick the metaphor that works for you. This is when the microbes are ramping up, but not to the point yet where you notice anything happening. When you notice the microbes working — usually indicated by millions of bubbles rising up along the container's walls or a sea-foam–like froth blowing up through the open top — this phase has ended because there is no more oxygen left for the yeasts to consume. It's the result of an unchecked population explosion. Because very little or no oxygen from the ambient air is getting back into the fermentation vessel, eventually all the oxygen is depleted.

How long this phase takes has everything to do with the yeasts and the temperature. For example, if you add sulfites, use a fast-starting commercial yeast, and the juice is at room temperature (upper 60s to mid-70s °F/16 to 24°C), you could see action in as little as a few hours of pitching the yeast and securing the airlock. If you use a less aggressive yeast strain and your juice is at a lower temperature (say the low to mid-50s°F/10s°C), don't stay up to catch the first bubbles because it might be a few days before they arrive. If you use wild yeasts at the same lower temperature, it can take 7 to 10 days before you notice anything happening. In any case, when the oxygen is depleted, the yeasts do something amazing: They switch food sources. Oxygen gone? Okay, now we feast on sugar!

PRIMARY FERMENTATION (YEASTS CONSUMING SUGAR)

If it's action you are looking for, this is the fermentation phase for you. First comes the foam — maybe. While it doesn't always happen, it often does. The foam, which can be anywhere from dirty snow–brown to pure white, appears on top of the must and pushes its way upward, through neck of the carboy, often arching upward before gravity pulls it down the sides of your carboy, causing a bit of a mess. It can be anywhere from a loose association of bubbles and fruit bits to a tightly packed foam similar to soft-serve ice cream. It doesn't always make it up and out of your carboy, but if it does, you'll want to remove your airlock to prevent even more of a mess.

If the cider is made with freshly pressed juice (and so has a lot of solids in it), we place a piece

THE NEED FOR NUTRIENTS

Yeasts are our partners in the cidermaking adventure and, well, they can be needy. They like it to be not too warm and not too cold, they like the cider pretty clear, and they like to have things like nitrogen available to snack on — especially in the early phases. The amount of nitrogen in your apple juice depends upon several things, including the cultivars used, the conditions in the orchard, and the maturity of the apples at harvesttime. Many, if not all, of these factors are outside your control and knowledge if you don't own the trees.

Testing for the yeast assimilable (or available) nitrogen (YAN) requires some lab equipment and the handling of formaldehyde, which we think is beyond the scope of most

cidermakers — including us. That's why we minimize the chances of having stressed-out yeasts, which is what happens when they don't have enough nutrients, by employing three tricks: we keep the temperature on the low end, we don't rush the fermentation process, and we choose commercial yeasts that have low nutritional demands. By doing this we have avoided the need to add nutrients.

What if your yeasts need to be fed? They will tell you they are stressed by producing sulfur, which you can easily detect when you give your cider a good sniff. If you detect sulfur, rack the cider to a sanitized carboy and add a yeast nutrient per the manufacturer's specifications.

of plastic wrap on top of the carboy opening in case it foams up. Sometimes we go straight to the airlock and watch for the foam. You will have some warning — it's not like your cider is clear at 11 a.m. and foam is blowing your airlock off by the end of lunch. The worst case is that you return for morning inspections to see your airlock full of brown foam. If the foam is coming up through the airlock, we take it off and retreat to placing a piece of plastic wrap over the carboy opening, unless it's too active for that. While the cider is actively foaming, it is very similar to having a 6-month-old baby in diapers. In other words, things are going to happen, and you can neither predict when nor stop them. But unlike diapers, cleanup is much easier. We simply wipe the foam away with a warm, soapy cloth. Once the foam stops attaining escape velocity, we put a clean airlock back on.

Depending upon the available sugar in the must, the type of yeast strains present, and the temperature, the bubbles will persist for days to weeks. During this phase the yeast population explodes, the sugars are consumed, and lots of carbon dioxide and alcohol are produced. It can be mesmerizing to watch.

This phase officially begins when you notice action in the form of bubbles, and it ends when you notice the lack of bubbles. If that seems less than precise, you can also track the specific gravity of your must starting at prefermentation, taking measurements every few days. The end of this phase is tied to when the yeasts are out of oxygen and pivot to consuming sugar instead, which is really when the alcohol begins to be produced

in great quantities. When the specific gravity has dropped to half to three-quarters of its fresh juice level, primary fermentation has ended. For example, if the fresh juice had a specific gravity of 1.060, primary fermentation is over when the specific gravity is in the range of 1.030 to 1.015. The downside to this method, in our opinion, is that you need to get into your cider, which opens up the possibility of introducing microbes that you don't necessarily want.

The threshold between primary and secondary fermentation is often the act of racking. Think of it as changing your apartment or house through a service that sweeps in and takes everything worth taking, leaving anything that you don't want to bring with you. Some feel that by pulling the cider off the spent yeasts, you give it a better chance to get on with the secondary fermentation's main goal of making the bulk of the alcohol. Others are less sure this step is really necessary, and instead, they choose to age on the lees, which you can read about on page 100. If you are in the beginning of your cidermaking journey, we suggest that you go ahead and rack between primary and secondary fermentations. Then, either by neglect one season or through curiosity, try skipping racking and see what your cider is like. If you like it, repeat. If not, keep on racking.

1. Cover the filled carboy mouth with a piece of plastic wrap to keep bugs out while you wait for active fermentation to start.

2. Primary fermentation begins when you notice the signs — often, gross lees and apple solids are blown off.

3. Keep a clean environment during the active blow-off phase by wiping down the sides of the carboy with a clean cloth.

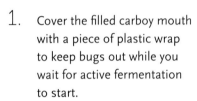

4. When the fermentation is less active, it is ready to be capped with an airlock.

5. Add water or a neutral distilled spirit to the airlock and check it periodically to make sure the liquid stays at the level specified on the airlock. When the primary fermentation has ended, rack the cider into a clean carboy.

WHAT ARE LEES?

Lees are the solids in the cider; they are generally defined as gross lees and fine lees. The gross lees are the large particles that are in the juice from the pressing — the bits of fruit, skin, seeds, leaves, or whatever else made it through the pressing bag. Depending on your pressing system, you could have a lot or a little of the gross lees. The fines are largely the microbes. In the beginning, there are multitudes, most of which don't live very long in the juice. As the juice turns to cider, the "biomass" that piles up is mostly yeasts and bacteria, some of which are alive, and many of which are spent at this point, and a little plant material.

The gross lees are more obvious in the early stages of the fermentation, and you will want to remove them. In most cases, they are naturally pushed out of the container during that first active fermentation, when the yeasts are working at full tilt and brown gunk comes spewing out of your carboy, bottle, or barrel. When you are cleaning the sides of that same container every morning, you are cleaning away the gross lees. That is good. You want them out; they do not offer any advantages to the fermentation. In fact, if the fermentation starts slowing down and you still see the brown gunk of the gross lees on top, top off the container with more juice. This will help the ferment push them out.

As the cider starts to clarify, you will see the fine lees collecting at the bottom of your fermenter, assuming you are fermenting in a clear carboy. If you are using stainless steel or a barrel you won't see them, but they are there. Standard practice is to remove these fines fairly early in the fermentation by racking the cider off them, leaving the dead yeasts and friends behind to be thrown out (though we suggest you consider aging methods that embrace the lees; see page 100 for more).

1. Place the carboy to be racked on an elevated surface.

2. Place an empty carboy on the surface below the full carboy.

3. Remove the airlock and insert a sanitized racking cane and siphon into the cider.

4. Drop the cylinder into the cider and the tube into the empty carboy. Pump the siphon two or three times to start the flow of cider.

5. Monitor the racking cane, being sure to keep it above the lees.

6. Cider from the upper carboy will not fully fill the second carboy due to the lees, so add additional finished cider so that the cider in the new carboy is within 2 inches of the top.

USE YOUR FIVE SENSES

Cidermaking is both a process and a journey. It is science, sure, but also art. It is facilitating a relationship between apples and microbes —a cooperative endeavor. How do you work with colleagues that are, well, microscopic? You watch, but also listen, smell, and taste what they have to say. Okay, even feel, but we will leave that to the troubleshooting on page 301.

In the first hours of fermentation, you will start watching the juice, waiting patiently for that first bubble or bit of foam collecting on the top. As the yeast becomes vigorous, the juice begins to get lively. At this point the fresh juice will no longer smell as sweet; instead it will smell mostly of carbon dioxide and yeast and possibly a little funky. You will see and hear the bubbles. In its most active stages, these bubbles will sound almost like popcorn. When the pops become further apart, you will know that the yeast is slowing down and has eaten most of the sugar. This is generally a good sign, but if you taste the cider at this point and there is still a lot of sweetness, it might also be an indicator that the yeast is stressed and may need nutrients.

While these early bubbles are creating carbon dioxide and pushing out the gross lees (page 91), you have another chance to understand the nuance in your cider. Before that daily wipe down (which you will read about in a moment), you can taste and smell those lees. We find that they are often bitter, and we are grateful that the fermentation is pushing this gunk out. Throughout the process, as you rack, bottle, and stir, smell, taste, and take notes, you will gain immeasurable insight. We invite you to stop, look, listen, smell, and taste your cider throughout the process.

SECONDARY FERMENTATION

This phase is not about fireworks. Gone is the explosive foaming, the sinuous rivers of continuous bubbles. This is the slow burn phase, when all the sugars will eventually be converted into alcohol — if a dry cider is what you're after. Your goal is to go slow and keep the temperature cool, especially if you are working with native wild yeasts. If you are making your cider from freshly harvested apples, this phase begins in late autumn to early winter and continues into early spring, or until things begin to warm up.

The cider should be at a consistent 55°F/13°C or less. We ferment and age our ciders in a room (our cave) built into the hillside just off the commercial kitchen in our farmhouse. When we built the kitchen and side-by-side caves, we used materials that gave us 12-inch-thick walls. We buried two sets of about a hundred feet of 4-inch pipes throughout the hillside to return the earth-cooled air to the caves. That means we get a consistent temperature of 60°F/16°C to 62°F/17°C, which is pretty nice for the early fermentation phases, as well as when we are aging ciders. It's a little warmer than ideal for the wild yeasts, so we manage this by opening windows at night and using fans to supercool the caves to get them to a lower temperature.

This phase concludes when the specific gravity has dropped to your target level or the malolactic fermentation has begun (we'll get to that next). If you don't want malolactic fermentation and/or are stopping fermentation before it completes to dryness, skip ahead to Aging (page 100).

Cold

At temperatures near freezing, yeasts stop metabolizing sugars. They don't die, they just slowly sink to the bottom and hibernate until things warm back up. By carefully racking the cider above these sleeping yeasts (or drawing off the bottom if using a conical fermenting vessel with a bottom valve), you can reduce the yeast population. Note that we said reduce, not eliminate. You could do this several times, but you won't remove all of the yeast, so it's best to place the cider in the refrigerator immediately after bottling and keep it there until you're ready to enjoy it. You will need a lot of refrigerator space, unless you are making very controlled batches and coordinating your consumption to keep it all balanced in the fridge shelves.

Heat

The last way to stop fermentation is by taking the temperature in the other direction and cooking these little guys to death, through a combination of temperature and time. The higher the temperature, the faster the carnage, and the longer you keep the temperature high, the more are killed. (However, if you are looking for a lightly probiotic drink, this will also take out any live bacteria.)

For those of you who have preserved your summer harvest by water-bath canning, you might be thinking this sounds similar, and you are right — it is. One option is to cap the bottles, fully immerse them in the water, raise the temperature to 149°F/65°C, and keep it there for 10 minutes.

However, if you go too long you get a cooked flavor.

To avoid that, and to avoid the possibility of an exploding bottle or two, we prefer to leave the bottles open and fill the water to a level that covers the cider line in the necks of the bottles. This way we can drop a sterilized instant-read thermometer down through the mouth of the bottle to get an accurate reading of the cider temperature. We think this makes it easier to heat the bottles to the minimum temperature and keep them there for the minimum period of time. Pull the bottles out of the water bath and cap as soon as you can safely handle them.

Heating cider in 149°F/65°C water to stop fermentation

AGING

Now that fermentation is complete, you might think it's time to finally put your cider in bottles and start the countdown until you can drink them. In fact, you could skip aging, move directly to bottling, and drink the still cider right now. You can also skip to the next step and add some forced carbonation (see page 108), in which case you could be enjoying a sparkling version in as little as a couple of days. You will likely have a better cider, however, if you age it for a few months.

Compared to the fermentation phase, this is a very quiet time and there is not much to look at. What you will notice, hopefully, is that over time your cider will clarify and become clearer, producing a growing layer of lees on the bottom of the carboy just when you thought there was nothing left to drop out. Since measuring the specific gravity won't tell you anything at this point, the only way you'll know this step is done is either when the cider has reached the level of clarity that you are looking for, or you lose patience and get on with the bottling. In the throes of spring on the farm, we used to move everything into bottles or kegs so that there was room for spring activities, while promising to do a better job of leaving them all to mature next time. The ciders we aged were usually the ones that got forgotten in the back of the shelf; by the time we found them, they were a clearer, more golden version of themselves. Now we find ourselves leaving many to age on purpose.

Aging on Lees

Another way to build unique and complex flavors in cider is to discover the hidden potential of the lees. The lees can be seen as an essential part of the terroir of place, as the lees contain all the bits that the apples may have picked up along the way — from leaves to microbes.

Usually cidermakers remove the lees to avoid hydrogen sulfide (page 303), a good reason for sure, but let's look at why we do this and what would happen if we turned this common practice upside down. What if lees are a good thing?

First, a clear juice can starve the yeast, and having some solids in your fermenting juice is important — they act as nutrients for the yeast. Sometimes you might add nutrients, or even packaged dried lees (basically yeast hulls) from other ferments, but they need to be kept

Cider aging on lees

WOOD-AGED CIDER

Let's face it, barrels are sexy — at least when they are filled with aging wine or, better, a great cider. They are also heavy, require care, and can be expensive unless you have a source for worn, played-out barrels from a local winemaker. If they are too well used, they likely will not impart much, if any, oakiness. Still, the oak notes in a cider can often take it one notch closer to heaven.

The type of wood, the amount of char (if any), and what the wood has come in contact with all play into what flavors will be introduced. Former bourbon and gin barrels are popular with some of our professional cidermaker friends because they are looking for hints of those previous occupants. Barrels often can only be used once in the distilling industry, so there is plenty of flavor left in the barrels for multiple seasons of cider.

If the idea of making at least 60 gallons of cider to fill a typical wine oak barrel (known as a hogshead) seems stressful, or the fact that its 600-pound weight means you will also need a forklift, don't worry. We don't have a forklift either, although it has been on Christopher's birthday wish list for years. Used American whiskey barrels are often narrower but still weigh in at over 50 gallons. Quarter casks are a little over 20 gallons and are used in the distilling business because they age things like whiskey more quickly since there is more wood surface area per gallon. There are also specialty barrels that can range all the way down to a gallon, but a word of warning: barrels are as much a work of art as a functional

Well-aged cider (right) is much clearer than fresh juice (left).

container, and the craftsmanship of the cooper who made that barrel will define your experience with it. We have one very cute 1-gallon oak barrel, which is beautiful to look at but terrible to actually use to age anything because in our dry climate, there is not enough liquid to keep the barrel hydrated — it evaporates in just a few months.

If you love the idea of a hint (or a heavy hint) of wood in your cider but don't have the space or lifting capacity for barrels, there are some easy and innovative options including oak chips, cubes, blocks, and spirals. For more thoughts on the subject, see Barrel Aging without the Barrel on page 103.

moving around. As fermentation takes place, the yeast cells settle at the bottom of the vessel, compacting each other as they do. This causes them stress and can result in more volatile acidity and sulfide production. However, when these yeasts (and bits) are kept in suspension during the third (maturation) stage of fermentation, they will actually help mature and add depth to the cider. This is done by stirring them up regularly, a practice called bâtonnage, which you will read about shortly.

Let's take that idea further. What if the lees were used to enhance cidermaking itself? Many small artisan cideries and home makers are borrowing from traditional cider and wine techniques to incorporate the lees in a process known as *sur lie*, which simply translates from French to "on the lees."

Why would you want to keep the lees around? The first reason is flavor. Wines and ciders that have been aged on their lees have umami, in other words that fifth flavor — deliciousness. If you don't believe us, look up the process for making heritage champagnes like Dom Pérignon. The bottles have been aged on their lees for 30 or more years, where the second fermentation happens with the lees in the bottle over the course of many years, until the lees are finally disgorged. This type of flavor building with lees can be used in cider as well, whether in the bottles with disgorgement or in the carboy with bâtonnage.

The lees also absorb oxygen, which will reduce browning and make the cider more stable, and, you guessed it, add more dimension to the cider's flavor. This helps maintain a slow and controlled oxidation during maturation.

But we have to come clean here. If your idea of delicious is a fresh and fruity cider, leaving cider on the lees may not be for you. You will get a complex marriage of acids and umami. Those of you who understand that umami is glutamic acid, which comes from breaking down proteins, might be thinking: *What?* It turns out the main nutritional component of the lees is protein, which is directly related to the extraordinary population of microorganisms — both alive and dead.

The dead yeasts start breaking down in a process called autolysis, which just means that they are falling apart and releasing all kinds of components — cell wall bits and what have you — that are interacting with the cider. The proteins are breaking down, as are the starches, which release a tiny bit of glucose and give the cider a little more sweetness. These released nutrients feed the malolactic bacteria (this is one of the places where you trade in fresh and fruity for aromatic complexity and flavor depth and length). Early in the malic fermentation, it can smell particularly funky. It is something to get through and come out the other end; when done well, things will settle down and become tasty. Because the lees

Wood spirals, sticks, or used barrel staves (called inner staves) are the easiest to use. In our experience, they also give you more control over the flavor. These can be found in a diverse variety of woods and multiple char levels, giving you a lot of flexibility in flavors. Tie a string to the end of a spiral and drop it into the carboy so that the spiral is fully submerged, tethering the string to the neck. The spiral of wood can be removed easily when the desired taste is achieved — no need for racking. They are reusable and will mellow over the course of subsequent batches.

We once made a cider with champagne yeast and 4 pounds of dark sugar in a 5-gallon carboy. Two months later it was clear dark amber with a high alcohol content. Christopher dropped in an oak spiral and aged it for 3 months. When we sampled it at bottling, this cider felt as close as you could get to a distilled spirit using just yeast. It had a hint of burn with bourbon notes. As soon as it was discovered by friends and family, it didn't last long. This process became our Bourbon Cider recipe, which you will find on page 278.

Wood chips or cubes are inexpensive; cubes have less surface area than chips. We tried a number of ways to use them — from making an oak tea to dropping them loose into the carboy. You can do this, but it's nearly impossible to fish them back out again without pouring out the cider. Therefore, one must plan to taste often and bottle when the flavor is on point. If you think you will be bottling as soon as you reach the right oakiness, don't worry about it and drop in the chips or cubes. If you think you would like to continue to age in the carboy after reaching the right oakiness, then fill a small mesh bag with the cubes and squeeze that into the carboy and down into the cider, using the string to tether the bag to the neck of the bottle and to pull it out when you are ready. (You will have to wrestle a little with the swollen wood.)

Chips are also used with an oak tea method. The chips are soaked in boiled water (like tea), and this tea is poured into the cider once it has fully steeped. We did not feel we had control over the flavor — if it was too strong there was no turning back, and if it was too weak it felt like a waste. Still, to give you a chance to try the oak tea method for yourself, we use it in our New England–Style Cider on page 191.

You shouldn't need to sanitize new chips, sticks, or cubes. If you are reusing them from a previous batch and you want to eliminate the chances of cross contamination, you could easily pasteurize them by placing them on a baking sheet in a 225°F/110°C oven for 10 minutes. You can also soak them in spirits both to sanitize them and impart the approximate aging tastes found in a used barrel, but be sure to taste often to achieve the flavor you are looking for. For example, we put 3 ounces of loose rum-soaked chips in a 1-gallon cider for 9 days. A few days in, Kirsten tasted it. It was a rum cocktail. After another few days, the cider tasted like it had been rolling around in a pirate's cask. The rummy chips had overpowered the cider, not leaving much nuance. You had to like rum to enjoy it. We suspect it could be used to mask a less-than-tasty cider.

are influenced by everything — from the soil of the orchard to the bees that pollinated the flowers to the sanitary conditions of the pressing and fermenting to choices made by the cidermaker — all lees are unique. As one would guess, the lees of wild ferments have a higher biodiversity, [9] which shows itself in the flavors as well as in their probiotic merits.

Finally, scientists are looking at ways that this "waste" product, which contains lots of fiber, protein, fatty acids, antioxidants, and probiotic microbes, can be used. In 1922, the creators of Vegemite, that salty Australian umami paste, figured out a process to capture B vitamins and other nutrients from the waste products of beer brewing, so the idea is nothing new. Lees are undervalued as of now, but that is changing.

Bâtonnage

It was Kirsten, who had almost zero knowledge of cidermaking at the time, who convinced Christopher to try *sur lie* with bâtonnage — the practice of stirring the lees back into cider. It went against all of Christopher's cidermaking training and instincts. Kirsten didn't have any prejudices against lees — to her they weren't something that could ruin cider — so when she read about the practice, she couldn't come up with a reason not to do it. Christopher agreed to sacrifice one carboy to her crazy notion. He carefully pulled it off the shelf, as he was accustomed to not disturbing the lees, and took a sample off the top to taste. It was what we might call a table cider — easy drinking, clear, crisp, and well, unchallenging (you could substitute the word *boring* here). The lees were compacted at the bottom. Kirsten took a stainless steel racking cane, dropped it all the way to the bottom, and stirred vigorously, kicking up the lees. The tornado of sediment worked its way up through the clear cider, making it dark and stormy. "We are disturbing the layer of dissolved oxygen. When we release that trapped oxygen and give the yeast motion, they are like little oxygen scavengers and will remove it, so that the cider won't oxidize. Isn't that cool?" she said. Christopher looked scandalized.

Kirsten kept stirring. The idea, she told him, was to keep things in suspension for a while. "We want as much of the surface area of the cider in contact with the lees for as long as possible," she said. She stopped stirring after a bit to smell the cider. "It still smells fine. Go ahead, smell — no hydrogen sulfide. It's gonna be great!" she exclaimed. Christopher didn't look so sure. "These ciders are already well into aging. If there was hydrogen sulfide we would have it by now. I read that if we haven't been getting it through the fermentation, we likely won't get it now with aging," Kirsten said. We topped off the carboy with some bottled cider and put it back.

A week later, the lees had settled again. We pulled out the carboy and repeated the routine. The flavor had changed already and not for the worse — in fact, there was much more flavor. In a spontaneous move, Christopher decided to hand five more carboys of cider to Kirsten to stir, as he still couldn't bring himself to do it. He watched each one, partly in fascination and partly in horror, as the cloud of sediment kicked up and swirled about. Tasting, stirring, and topping off these carboys became part of our weekly routine for the next 9 months, the time needed to get the fullness of the yeast flavors.

The yeast cells autolyze at the end of fermentation, but they do so slowly. It is when autolysis really kicks up, after many months, that the amino proteins are released. These proteins are found in the cells of the yeast walls and are known to affect the mouthfeel of cider (and wine).

The flavor of every cider we aged on lees was enhanced greatly. We got a wide range of flavors — everything from a buttery chardonnay cider to a deeply multifaceted farmhouse.

Fining

During the aging process, it's possible for the cider to develop a haze that consists of tiny particles that remain suspended in the cider and refuse to clump and drop to the bottom. To force them to clump, we need to break their electric charge through positively charged proteins, like special types of gelatin, and negatively charged particles, like clay or silica. This is traditionally done by mixing up portions of each and using samples of your cider to experiment with different ratios until you find the clarity you are looking for. Then you scale up that ratio to the size of your batch and introduce it.

We could describe the lab experiment for you, but honestly, we have moved to something simpler and more convenient: two-stage fining agents. You can find them in brewing supply shops and on the

Using the bâtonnage method to incorporate lees

Internet; they come in two connected pouches that are often advertised as a kit and are available for a few dollars. Follow the manufacturer's directions, which is basically to stir in one packet (the negatively charged silica) and wait a day, then stir in the positively charged gelatin substance. Wait another day and your cider should be much clearer. Either rack to another storage container if you have more aging to do, or rack directly into bottles if the cider is ready.

BOTTLING

The racking skill you learned when transferring the cider between the primary and secondary fermentation vessels will come in handy for bottling because you are basically doing the same thing. Fill sanitized bottles to within 1 inch or less of the top, then seal them by closing the bail, corking, or capping. If you want a still cider at the current sweetness, then you're done. If you want bubbles or a sweeter cider, then you need to take some steps to add them.

Back-Sweetening

Your cider has finished dry (specific gravity of 1.000 or less) but you would like your cider to have a little, or more than a little, sweetness. If you just put sugar in the cider, like we do later in this chapter when we add priming sugar to bottle condition, any remaining yeasts will happily wake up and consume the sugar that you want for yourself. That means it's either them or you. If you want that sugar, you need to take the yeasts out. There are a few ways to do this, which we have already covered when we discussed stopping the fermentation party on page 97. It's important to realize that just because your cider is dry, it doesn't mean no yeasts are present — at least not right away. The longer your cider has aged after fermentation is complete, the fewer yeasts remain.

In our experience, once fermentation is complete (specific gravity of 1.000 or below) and the cider has aged for at least 3 months in a cool environment (50°F/10°C to 55°F/13°C), it is ready for bottling. To make sure, a week before we plan to back-sweeten and bottle, we draw out a pint with a wine thief and pour it into a freshly emptied disposable plastic water bottle. We cap that and leave it on the counter in the kitchen for a week. If any pressure is created, you will notice the bottle swelling or you will hear the hiss of carbon dioxide escaping when you open the cap. That likely means you still have active yeasts and you need to let it age another month before checking again. Otherwise it's time to add some sweetness.

The sugars you could add fall into two camps: those that yeasts can't ferment (just in case there are still some in your cider) and sugars that yeasts and humans enjoy. That's probably showing our hand because we are not fans of unfermentable sugar products, be they artificial ones like Equal or Splenda or natural ones like stevia. Ultimately, if you choose to consume these products on a daily basis, then feel free to include them in your ciders. Otherwise, if they are not part of your daily diet, we suggest forgoing them and moving on to more natural choices.

We prefer adding sweetness using honey or raw sugar. Mix equal amounts of the sugar with boiling water and stir to completely dissolve in the water. For honey, allow the boiled water to cool to 100°F/38°C. Start with 1 cup of sweetener and 1 cup of boiling water for a 3-gallon carboy. Add it to the carboy and swirl to combine, then use a wine thief to draw out a sample to taste. Is it where you want? If not, repeat until it is.

1. Siphon the cider from the carboy into sanitized bottles.

2. Fill the bottles to within ½ to 1 inch of the top.

3. Cap the bottles by drawing down on the levers to engage the capper and secure the cap.

Carbonation

Want a nice pop when you open your cider? Something between an Indy 500–winning driver shaking the giant champagne bottle bubbles and such an imperceivably small fizzle that you could be opening tap water bottled that morning? There are two options for getting bubbles, or more precisely carbon dioxide, into your cider.

FORCED CARBONATION

This is just what it sounds like. We are forcing carbon dioxide into a still cider in the bottle, though technically it usually happens in a metal tank, keg, or growler. Larger producers chill their finished cider in stainless steel tanks equipped with cooling jackets. Once the cider is very cold, they inject carbon dioxide into the entire batch, then bottle it while chilled with the dissolved carbon dioxide. These tanks start in the hundreds of gallons and thousands of dollars and go up from there, so they're not an option for most of us.

Enter the ubiquitous 5-gallon (19 liter) Cornelius keg, also known as "corny" kegs. They are nearly indestructible and are used in both the soda and beer industries, so it is pretty easy to get a used one in good shape. It is a metal cylinder with a removable lid/handle combo that has a couple of ports, one with a hose that extends to the bottom of the tank and another that is shorter and extends near the top. The liquid is usually dispensed through the port with the long hose while pressurized gas is connected to the port with the shorter hose, keeping it at a constant pressure in the tank. We reverse these ports by connecting the carbon dioxide tank to the long-hose dispensing side so that the carbon dioxide enters from the bottom of the tank and dissolves in the cold cider as it floats up. See the step-by-step process on pages 110–11.

SQUARE-SIDED AND NOVELTY BOTTLE CONSIDERATIONS

Bale-style bottles are popular for water and decorative uses and can be found in varying degrees of quality from retail stores. We have discovered that some of these have gaskets that don't fully seal, which can cause leakage or a slow deterioration of flavor due to the small air exchange. Some of these bottles are not manufactured to handle the pressure that can build in live fermented carbonated beverages. Be aware that square-sided bottles are weaker than round bottles due to weak points created by the shape.

After you have carbonated the cider in the corny keg, you can either fill bottles and seal them or you can just enjoy your cider right out of the keg. As you drink from the keg it will lose its pressure, so to maintain bubbles over several days or weeks you might need to reattach the carbon dioxide tank to the shorter-hose input side and add gas.

The corny keg works nicely, but it's big and not everyone can afford the space to refrigerate it with a kegerator setup, nor do some of us need that much cider readily available at just the pull of a tap. We went in search of something smaller that required less equipment and found the uKeg pressurized growler from GrowlerWerks, based in Portland, Oregon. It's basically the same concept as the corny keg, except in a ½-gallon or 1-gallon size. The pressurization comes from a small carbon dioxide cartridge that screws into the lid.

We have found that while the uKeg sits in the fridge for the first couple of days, it helps to pull it out once a day and give it a gentle swirl/shake for a minute. After a week or so the carbonation diminishes as the cider gets low in the growler.

The uKeg isn't a great solution for filling bottles, since you can only do a gallon at a time, but the portability of the growler means you can skip the whole bottling thing altogether and go from finished still cider to carbonated cider in a couple of days.

1. Following the sanitizer manufacturer's directions, mix 3 gallons' worth of sanitizer in a 5-gallon bucket.

2. Disassemble the keg components (dip tube, O-rings, keg lid, connects, and poppets) and inspect. Replace any O-rings and place in the 5-gallon bucket of sanitizer, along with the autosiphon and tubing.

3. While the components are sanitizing, use hot water to rinse out the sides and bottom of the keg.

4. After 3 minutes of full contact with the sanitizer, place each component on a clean towel. Pour the sanitizer into the keg.

5. Reassemble the tubes and connects and reattach. Attach the lid securely, then swirl and shake to ensure contact throughout the keg. Let it sit for 3 minutes, turn upside down, and wait for another 3 minutes.

6. Open the keg and discard the sanitizer.

7. Siphon 5 gallons of refrigerated still cider into the 5-gallon Cornelius keg.

8. Inspect the keg lid and make sure the dip tubes and connectors are securely attached. Fit the lid on the cylinder and make sure the gaskets are seated properly.

continued on next page

9. Purge the oxygen in the headspace by connecting the carbon dioxide tank line, opening the gas valve, then opening the regulator valve to 30 PSI. Wait for 1 minute and pull the pressure relief valve (if ball-lock style) or remove the gas line and depress the poppet (pin-lock style) for three short pulls to purge the oxygen.

10. **Fast option:** To speed the carbon dioxide saturation in the cider, increase the surface area between the cider and the carbon dioxide by laying the keg on its side and rocking it back and forth for 3 minutes. Reduce the pressure to 10 PSI and give the keg a quick pressure relief. Refrigerate for at least 3 hours.

11. **Slow option:** Reduce the pressure to 15 PSI and give the keg a quick pressure relief. Refrigerate for 2 to 3 days.

12. Connect the spigot attachment line and pull a sample to taste. If it has the carbonation you are after, you are done, otherwise adjust the pressure to your desired carbonation.

1. Clean and assemble the growler, cap, unused carbon dioxide cartridge, and cartridge sleeve.

2. Place the cartridge with the narrow end facing the opening of the sleeve. The lid gauge should be in the off position.

3. Tighten the sleeve onto the cap. When you feel resistance, it is making contact with the carbon dioxide cartridge so twist quickly and firmly until it is tight.

4. Pour the cider into the growler until it's completely full.

continued on next page

5. Screw on the lid until tight, then dial the pressure gauge to the desired level of carbonation. Make sure the dispensing handle is locked.

6. Give the growler a few gentle shakes, then place in the refrigerator for 2 to 3 days.

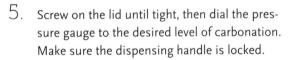

7. Unlock the tap and enjoy carbonated cider that was still only a few days before.

BOTTLE CONDITIONING

The second way to get bubbles into cider is to bottle condition, which is the process of setting up an environment in the bottle that has just the right amount of yeasts and unfermented sugars so they continue to do what they do best, producing a little more alcohol and a little more carbon dioxide, which is what we are after. This shouldn't be confused with another technique called back-sweetening, where we add sugar at the time of bottling — not for the yeasts to ferment but for us to enjoy. With bottle conditioning we are not after more sugar but more bubbles.

The amount of bubbles and, more importantly, the pressure that is built up in the bottle are determined by the amount of sugar you ferment, the yeasts that are present, how well the yeasts convert the sugar to alcohol (attenuation percentage), and the time and temperature that the yeasts have had to do their work.

We will cover four options for bottle conditioning. The first, *pétillant naturel*, bottles the cider while the yeasts are still converting sugar to alcohol. The second and third methods wait until all the initial sugars have been consumed, then introduce a controlled amount of additional sugar to the bottle. The third option also reduces the risk that the yeast might not be strong enough to keep fermenting by adding new commercial yeast. All three of these methods will produce sediment in the bottle and perhaps result in a cloudy cider. The last method also clarifies the cider through the riddling and disgorging processes.

Which to use? If you are new to cidermaking, we suggest starting with option 3 because it doesn't require any special equipment and is the most controlled. After you get a few good batches under your belt, you might move to option 2, gaining a better feel for your yeasts. From there, options 1 and 4 are for those looking to up their game; they require paying a lot more attention to what you are doing and when.

Option 1: Semi-Sparkling (*Pétillant Naturel* or *Pét-Nat*)

Although this process has been around for hundreds of years (it's thought to date back to French monks in the sixteenth century),[10] it is not for the faint of heart because it's really difficult to accurately predict the final carbonation level in the bottle. That's because you are bottling a work in progress, and if it's bottled too early, there may be too much carbon dioxide left to be created — that's when you risk a "bottle bomb." If done too late in the process, you get a "bottle dud," meaning you end up with still cider when you were expecting something more effervescent.

As the name suggests, we are trying for a natural semi-sparkling cider. For safety reasons, we suggest that you first shoot for something between still and semi-sparkling by bottling when the specific gravity reaches 1.002. Wait at least 2 months before opening and evaluating the level of carbonation. Once you feel more confident, try bottling at a specific gravity of 1.004 and wait for at least 3 months before opening and evaluating the carbonation and, of course, the taste.

Option 2: Adding Priming Sugar

This option gives the yeasts that have consumed all the sugar in your cider a bit more to work on in the bottle. Even if there haven't been any bubbles in your airlock for weeks, it's likely there are still viable yeasts in your cider just waiting. It's possible, however, that with this method the yeasts are not at a sufficient number or they don't have enough nutrients left to finish the added sugar to complete dryness. This can be a good thing if you are looking for a slightly sweet and effervescent finished cider. If the yeasts are sufficient and the cider is finished to full dryness, this should produce a semi-sparkling cider.

There are several ways to add priming sugar at bottling time. You can add a measurement of sweetener — either sugar or carbonation drops (sugar and corn syrup formed into drops) — to each bottle before adding the cider or you can make your own simple syrup and add that to the carboy of juice. The problem Christopher has with adding sweetener to individual bottles is that when bottling a lot of batches, he gets a little bleary after a while (could be all the mandatory tasting that this requires) and occasionally gives a bottle or two a double dose, which can provide an effervescent surprise come opening time.

Option 3: Adding Priming Sugar and Yeast

This option doesn't take any chances and uses a dry cider with a specific gravity of 1.000 or below and adds a controlled amount of sugar and champagne-style yeast to produce the desired carbonation.

Option 4: Riddling and Disgorgement

This is called *méthode champenoise* and it differs from the previous option in two ways: First, we want a sparkling cider, so we are doubling the amount of sugar. Second, we are going to try to remove all the spent yeast and sediment from the bottle so that we end up with a clear and sparkling cider. This time we handle the bottles weekly in a process called riddling, which concentrates the

LEVELS OF CIDER BUBBLES

- **Still (*perlant* in French).** Minimum carbonation; it likely won't produce foam when poured, but you may taste subtle bubbles as you drink it.

- **Semi-sparkling (*pétillant* in French).** Noticeable carbonation, with light foam when poured and a steady stream of bubbles rising up.

- **Sparkling (*bouché* or *mousseux* in French).** Strong carbonation, with a heavy foam when poured and strong bubbles when drinking.

lees in the neck of the bottle, and then we remove these lees through a process called disgorgement. We only use sparkling wine–style bottles that are designed to handle the pressure created by higher carbonation. Your choices here are either bottles that take crown caps or those that are the cork-in-a-cage style. We use bottles that have bail-style tops but also take crown caps, which we think are perfect for this process.

To concentrate the lees in the top portion of the bottle, we need to rotate the bottles weekly from a starting position of about 10 degrees (nearly horizontal to the floor), ending at 90 degrees from the floor (perfectly upside down). Traditionally, a riddling rack is used for this purpose. You can purchase one or, if you have access to woodworking tools and are handy, try making one. Just remember there will be a lot of weight and cider hanging in the balance. We prefer using a simpler method that we read about in the Farnum Hill Ciders' book *Apples to Cider*. Fill a large milk crate with bottles in the upside-down position, tip the crate on its side, and position a 2 × 4 board under one side of the crate so that the bottles are nearly horizontal. Store the bottles in a cool place (52°F/11°C to 57°F/14°C). With a nonpermanent marker or using thin strips of tape, mark the bottom of each bottle facing in the same direction.

Set a weekly alarm. Every week, give each bottle a quarter turn (use your markings to align all bottles together) and a light thump with your finger to gently encourage the yeasts to begin sliding down the neck of the bottle. Carefully add another board, thereby increasing the angle of all the bottles in the crate. Repeat weekly until the crate is sitting fully upright on the floor and the bottles are upside down. Let them sit in this position for 2 weeks.

Transfer the bottles in the upside-down position to a refrigerator turned down to 33°F/1°C or a chest freezer turned up to the same 33°F and chill for 24 hours. Now, be warned that the latter can be tricky if your freezer, like ours, does not allow for a temperature above 32°F/0°C. We usually pay close attention and turn the freezer on and off as needed, but one time there was a lot going on and we forgot to check it. When we did, we found that the bottles had frozen and exploded, so all the bottoms were at different levels, like a desert landscape with mesas and formations. When we pulled out the first bottle, the yeast plug was pushing its frozen self from the bottle, with the bottle cap stuck to the top of the plug. This situation can be remedied with a temperature controller that bypasses the unit's thermostat by cutting power when the temperature gets below the desired temperature.

The next part, disgorging, is going to be messy, so it's best done in a garage, on a driveway, or where it's easy to wash up afterward. Once you start you want to go through all the bottles at once, so have everything ready. You'll need a bottle opener, new crown caps and a capper, a funnel, and plenty of dry cider to top off the bottles. Just to be safe you should wear safety goggles and gloves.

Keep the bottles upside down until you're ready to open them, then turn them right side up and remove the cap. The pressure should push the plug of yeast out, and sometimes a bit of cider comes out too. As soon as the stream has calmed down, insert the funnel and pour in the new dry cider, and recap. The first few bottles are going to feel awkward, but you will soon get the hang of it. After you have disgorged all the bottles, wash them off and store them, this time in the upright position, until you are ready to drink them.

Oops! This is what happens when the freezer's thermostat is off and you wake to frozen broken bottles of cider.

Frozen yeast plug

5. Fill a milk crate with the bottles in the upside-down position.

6. With a permanent marker, chalk, or tape, mark a line on the bottom of each bottle all going in the same direction. This will help you know you've turned the bottles a quarter turn.

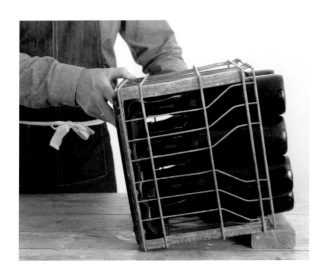

7. Tip the crate on its side and carefully add a board under one side to tilt the crate as near to horizontal as you can.

8. Each week turn all the bottles a quarter turn in the same direction (the lines should be aligned) and add one board until the crate is sitting flat on the floor and is no longer supported on one side by boards.

continued on next page

9. Place the crate in the refrigerator or in a chest freezer and cool until near freezing.

10. Revert the bottle while removing the crown cap. The pressure from inside the bottle will push out the yeast plug.

11. When the cider has calmed down, pour new dry cider into the bottle, filling it to within ½ to 1 inch of the top.

12. Recap the bottle and wash off the outside. Store upright where you store your other ciders, preferably in a cool spot out of direct sunlight.

Brady Jacobson of Mt. Hood Organic Farms & Nate Ready of Hiyu Wine Farm

Floréal cider is probably one of the best ciders that you have never heard of, unless you are a sommelier in Paris, London, or one of the larger cities in the United States. That's because there is zero advertising, only the word of mouth that comes from trust in winemaker-turned-cidermaker Nate Ready. In turn, Nate would only launch a cider after finding the best and most diverse apples anywhere — those of a relatively nearby biodynamic family farm run by the Jacobsons. It's a partnership that was years in the making, but it makes perfect sense — and amazing cider.

Christopher first tasted Floréal during an all-day workshop called "Exploring Commercial Craft Hard Cider & Perry Production" put on by Washington State University. At the tasting after the workshop,

Christopher was first in line when the 750 mL bottle of Floréal was opened. It was light gold and hazy, with a little yeast and fruit on the nose. Christopher later described the taste to Kirsten as "welcoming and comforting, like meeting someone for the first time and not wanting the conversation to end."

Brady Jacobson and her husband, John, run the Mount Hood–area farm that grew the apples for Floréal. They bought the abandoned farm in 1981, but it took a few years to transition it to organic certification, when they became Oregon's first commercial organic fruit producer. There were no packing houses for organic produce, so they borrowed a wood mill and felled trees to build their own packing and cold storage facilities. When they were ready to sell their apples, they found that marketing channels

for organic fruit were not yet in place, so they pioneered that as well.

When we got on the phone with Brady, one of the first things she told us was "as a small farmer you can never coast." Circumstances are always changing, and so must your approach. The idea to make cider began to germinate in Brady's mind about 15 years ago as the fresh organic fruit market continued to evolve into a place that was more about price than quality or relationships. Many years later, she met Nate Ready at a local farmers' market. Nate was visiting the area and complimented her on the diversity and quality of her apples. He asked her if she had ever considered making cider. She had for years but had not met the right cidermaker who understood the biodynamic philosophy and would bring that out in the cider. Nate did, though it would be several more years before the partnership would form, eventually producing an amazing product.

Floréal's first vintage was 2014, magnums of which were handled as a true *méthode champenoise* and riddled. Their cidermaking process is wild: they use apples, 35 varietals and counting, plus perry pears and quinces, fermented traditionally and slowly. They sweat the apples for 1 month before milling. The pomace is aged before pressing, then it's fermented and aged in seasoned barrels until the next year's harvest, when the new season's juice is available. At that time, the cider is bottle conditioned using the new harvest juice and aged in the bottles for months (or years, in the case of the 2016 vintage). Their ciders are available at their tasting room or online at Hiyu Wine Farm.

Brady is thinking about making cider vinegar and brandy next. All three of their daughters are now involved in the business, so it's not completely her decision anymore, she said. "I'm tired of wearing all the hats," she told us, and is looking forward to the younger people's energy and where it will take them.

PART TWO

RECIPES:
LET THE FERMENTATION BEGIN

CHAPTER 4

WILD CIDERS

There is beauty in simplicity. This chapter covers how to ferment apples (and other fruits) to produce ciders in the age-old way — just apples and the microbes they come with. The nuance comes from variations in process and aging styles. We explain different methods and introduce you to some makers who create beautiful place-based ciders in just these ways.

Remember: this is your cider adventure. There is no judgment. We don't want anyone to miss out because they don't want to go wild. If you prefer to use commercial yeast, please do. Just give the wild yeasts a few days to do their thing before adding their commercial cousins. While the recipes in this chapter give instructions for making the ciders wild-style, every one of them can be made using a commercial yeast and vice versa for the cultivated ciders in the next chapter, as long as the juice has wild yeasts, meaning it's not pasteurized or chemically treated.

AN ARGUMENT FOR WILD CIDERS

Bread bakers, winemakers, and cidermakers all rely upon yeasts to convert sugars to carbon dioxide and ethanol. Packaged pure yeast strains are a fairly recent addition to the human-yeast relationship. It wasn't until the late nineteenth century that industrialized yeast started to be grown for bakers and brewers, following Louis Pasteur's work of isolating individual yeast strains, most often different strains of *Saccharomyces cerevisiae*, who are the rock stars from breads to beers. In the case of bread, that ethanol is baked off in the oven and the gas is trapped by the gluten to provide the rise we all love. For fruit wines like cider, these strains are touted for their ability to convert sugar to ethanol in a range of temperatures and levels, and in the presence of small amounts of sulfites and rising levels of ethanol. They are predictable. They are like the rock bands of our youth. We don't want to hear new things from them; we want to hear those few favorite songs we remember that make us feel younger. The same could be said for the strains of *S. cerevisiae*; they have been isolated and bred because they do one or two things very well. They show up and do those things every time.

For every rock star, there are hundreds or thousands of singer-songwriters pouring their hearts out in coffee shops, bars, and summer festivals. Wild yeasts are those singer-songwriters of the microbe world. They are unique and do their own thing — unless they are a cover band, but that's another story. The wild yeasts that show up on and in an apple are the product of the orchard environment, including what other plants were in bloom nearby, what pollinators visited their blossoms, what nutrients were in the soil, how much it rained or didn't, what variety the apples are, and how the apples were harvested and stored. That microbiome fingerprint is unique to the tree and probably right down to the apple, and it seems to be a great shame to start the cidermaking process by dropping in some crushed Campden tablets and napalming everything, don't you think?

INDIGENOUS YEASTS AND BACTERIA

YEASTS, and more broadly, fungi, which are both yeasts and molds, are naturally on apples, and even on the blossoms that turn into apples. The nectar in the flower blossoms attracts pollinators like honeybees and bumblebees. Less than an eighth of a teaspoon of nectar has been found to contain up to four hundred million cells of yeast, and the number of bee visits are believed to be a strong factor in the variety and concentration of yeasts found in the apple.[11] This research shows how yeast gets inside an apple (basically, bumblebee butts), but what about the outside of the apple?

The apple's skin is a natural catchment for all kinds of bacteria — beneficial, benign, and pathogenic — and yeast. However, in a study in Spain, none of the yeast species associated with spontaneous fermentation were identified on cider apple skins.[12] A study of microbiota on the surface of apples and pears in Italy found a heterogenous mix of yeast, bacteria, and mold across the surfaces, with higher concentrations at the basal and stem ends of the fruit.[13]

While the yeast genus *Saccharomyces* is responsible for the final stages of alcoholic fermentation, it is surprisingly absent from the apple. A study in Spain identified non-*Saccharomyces* genera like *Kloeckera, Candida, Pichia, Hansenula, Hanseniaspora,* and *Metschnikowia* in apple must.[14] Scientists tracked the microbiota changes from apple must to final cider of two processing types and over 2 years of production. What they found was a varying mix of yeast species and density ratios in the apple must, which was attributed to geographic location (both fruit source and processing operation), climatic conditions experienced during the apple's growth and maturity, as well as the processing techniques employed. When the freshly ground apples were pressed over a period of 3 days, the *Saccharomyces* species was given an opportunity to not only establish itself but begin its ultimate dominance of the fermentation process, which began on the fourth day. Nature is full of collaboration and competition.

BACTERIA are present from blossom to bottle. Some families of bacteria, like *Lactobacillae* and *Acetobacteraceae*, occur in the flowers, the fruit, and in the cider. Others, like *Leuconostocaceae, Sporolactobacilliaceae,* and *Sphingomonadaceae,* only occur in the cider. One family, *Enterobacteriaceae,* is in the flowers and apples but it's not found in the cider. The number and types of bacteria on the apples before milling have been shown to be highly affected by how the apples were harvested and stored, with ground falls and bruised or rotting apples dramatically increasing the bacteria present in the apples before processing.[15]

In one study in Spain,[16] the evolution of yeast and acetic and lactic acid was tracked in four batches over 150 days, utilizing traditional methods at three different cider facilities. The fermentation tanks were traditional wooden 10,000-liter (2,642-gallon) containers that took between 2 and 9 days to fill with cider. Samples were drawn from the cider in the middle of the tanks. While the tanks were filling, the acetic acid populations were at their highest, but they dropped considerably as soon as the tank was full (and presumably oxygen was no longer accessible). Lactic acid bacteria remained steady or grew slightly throughout the fermentation process. That's important, as they are one of the key bacteria involved in malolactic fermentation; however, too much and they can produce spoilage compounds.[17]

Yes, wild ciders are not as predictable and safe as those made with commercial yeasts, and if you are betting your future on your cidery's success in producing reliable flavors batch after batch, we understand the desire to reduce your risk. Still, there are many commercial cidermakers making thousands of gallons of wild cider that are pretty consistent year after year, so it can be done. Perhaps most importantly, it's a very creative way to go about making cider. Instead of controlling everything, you are working with the microbe team already there, helping them to fully express what they were meant to do. What could be better?

WILD YEAST PRIMER

There are six main genera of yeast species on and in apples, waiting to ferment the apples' sugars into ethanol: *Candida, Kloeckera, Metschnikowia, Pichia, Torulopsis,* and *Rhodotorula.* The first two, *Candida* and *Kloeckera,* do the majority of the fermentation, but their tolerance for ethanol only reaches the ABV level of 5 or 6 percent, according to some studies. Ethanol acts as a preservative for cider, so low-ABV ciders (below 5.5 percent ABV) do run the risk of spoiling in the bottle. You can plan to drink these low-ABV ciders when they are younger. If you want to age them, you can reduce this risk of spoilage by addressing other factors that also contribute to your cider's preservation: adequate acidity (low enough pH) and the use of sulfur dioxide at the end of fermentation to eliminate all the microbes in your cider before bottling. That said, we hardly ever check the pH, and we rarely do the last one if wild yeasts have played a role because it's a live food at that point. We also rarely have a problem — most of our wild ciders age beautifully. If you have a batch that goes south — meaning it was nice at bottling time but it matured into a bottle of unpleasantness — it's time to go a level deeper the next time around and pay attention to things like the final ABV, pH, and acidity.

Generally, the natural sugars in apples will produce a cider that is 7 to 8 percent ABV if given a little help from the *Saccharomyces* species,

BASQUE CIDER: CHANGING POPULATIONS

When we visited the Basque cider house of Zapiain, Miguel Zapiain explained that after nearly 400 years of making wild cider, they started pitching champagne yeast over a decade ago because EU regulations required a change in practices. He said that the pitched yeasts controlled the primary and secondary fermentation, but as the cultivated yeast dies back, the native yeasts are ready to continue the ferment, adding their own flavor imprint. And it works: the flavors are unique and on point to the region. This, of course, goes against conventional wisdom, but the microbes work things in ways that we have yet to understand.

which can join the party in one of two ways. The first way it can be introduced is through your equipment, which can harbor the spores from previous cider batches. Our wild ciders regularly ferment dry, meaning they are consistently 7 to 9 percent ABV. This might be due to our wooden cider press or our barrels, buckets, or carboys carrying the *Saccharomyces* species of past ciders. The second way is to pitch it with commercial yeasts, which you can absolutely do without first laying waste to the wild yeasts with a carpet bomb of sulfur dioxide. We have found that if we want the final ABV to be a little higher than the wild yeasts can accomplish, or if they are stuck in the primary fermentation phase, it helps to pitch commercial yeasts at the end of the primary fermentation.

Finally, we just want to note that the wild-yeast-cultivated ciders we make using commercial, pasteurized juice as a base, fully ferment to dry with ABVs (depending upon the specific gravity of the base juice) of 7 to 8 percent. They do so even though neither of the two methods we just mentioned for introducing *Saccharomyces* species is a factor. We didn't pitch commercial yeast, and we fermented directly in the 1-gallon glass jugs that the juice came in (no wood presses or scratched plastic buckets that might be harboring those yeasts were used). The only yeasts that we know of came from the flowers that we added to provide the yeasts for fermentation. So, what explains this? We don't know, and we might be beyond the edge of current research — at least we couldn't find a study of this nature out there. For now, we chalk it up to the power of the flower.

HOW YEAST POPULATIONS CHANGE IN THE FERMENTATION PROCESS

A study in a traditional cidery in Ireland looked at 12 cider batches over a single season.[18] They divided the alcoholic fermentation into phases, and found that different yeasts were active during different phases of the wild fermentation.

FRUIT YEAST PHASE. This first phase was dominated by *Kloeckera/Hanseniaspora uvarum*–type yeasts that were found to be indigenous to the cider apples. They dominated the apple must (over 80 percent on average, but it varied), with some batches having a surprising concentration of *Saccharomyces cerevisiae* that were already present. It was found that the bruised fruit had the highest concentrations of *S. cerevisiae*. The other source of yeast that fed this fruit yeast phase was traced back to the apple wash water, with higher concentrations coming from water that was less frequently changed. Both *Kloeckera/Hanseniaspora*– and *S. cerevisiae*–type yeasts showed exponential growth after an initial lag time of 24 hours.

ALCOHOLIC FERMENTATION PHASE. As the alcohol level rises to 4 percent at about day 3, the *Kloeckera/Hanseniaspora* begin to lose steam and are quickly dominated by *S. cerevisiae*, which continue to dominate this phase until their nutrients, including sugar, are depleted.

MATURATION PHASE. In the final phase, *S. cerevisiae* runs out of fuel and is replaced by *Dekkera*– and *Brettanomyces*–type yeasts, which were present in the cider facility outside the fermentation season. At day 12, squarely in the alcoholic fermentation phase, the *Dekkera/Brettanomyces* yeasts were at concentrations of about 12 percent of total yeast, but by day 22, right in the middle of the maturation phase, they had risen to 90 percent.

One final point about this Irish study that we find important is that as the season progressed, the population handoffs outlined above didn't change radically but the timing of them did. What took 12 days to ferment in very high temperatures of 93°F/34°C to 97°F/36°C in September took a full 40 days at lower temperatures of 50°F/10°C to 54°F/12°C, with some dips to near freezing, in December.

Jonathan Carr & Nicole Blum of Carr's Ciderhouse

With the resurgence of cider in the United States has come a resurgence of cidermakers, and some of them have taken on the additional challenge of growing their own apples in a sustainable and flavorful way. It is a lot to master. And yet, spend a little time with Jonathan Carr and Nicole Blum and you completely believe it is not only possible but it is also the future we hope for.

We met Jonathan and Nicole in 2018 at the Mother Earth News Fair in Pennsylvania, where they were leading cider workshops and promoting their brand-new book *Ciderhouse Cookbook: 127 Recipes That Celebrate the Sweet, Tart, Tangy Flavors of Apple Cider*. Like us, they are a husband-and-wife team

trying to make a go of it on a farm by diversifying their revenue streams while not losing sight of the bigger drives in their lives.

They were seasoned market gardeners when they bought an abandoned orchard on the northwest slope of Mount Warner, near Hadley, Massachusetts. It would take years of hard work to remove the swaths of poison ivy, bittersweet, wild grapes, and Virginia creeper and to prepare the soil to the point that new tree blocks could be planted. Like master chess players, they are deep thinkers who can explain why the things they are doing in the fields now are in preparation for things they want to be doing years from now. They are several years into grant-funded research to create new varietals that

are the best of the old varietals — like one of their favorites, Golden Russet, as well as newer, more disease-tolerant varietals. Jonathan says things are coming along well with the research and they might have some promising varietals; they will know for sure in about a decade.

Carr's Ciderhouse produces 2,000 gallons of cider a year, and that's their sweet spot. As Nicole explained, to move beyond that, they would need to bring in a lot of inputs from afar — not just purchased juice, but also the sulfites and the filter technology purchased juice often requires, as well as more labor. As she explained to us, they think about their carbon footprint and consciously challenge themselves to stay within the boundaries of what they can source locally.

They are also not above challenging their devoted fans. As Jonathan explained, "You can make what you know people like, but you can also challenge them with flavors that are unfamiliar or that have been lost." For example, many of their new customers had never tasted a truly dry natural cider before. All of their ciders are now wild fermented and bulk aged before bottling. "We generally try to disturb our cider as little as possible," Jonathan told us. He has developed a unique bâtonnage technique for when he wants to utilize the lees to make a cider a little more interesting. Ultimately, their style could be summed up as long and low — a long fermentation at low temperatures to favor the bottom-fermenting natural yeasts and discourage any bacterial competition.

Wild Cider

Cider made from wild yeasts is not as unpredictable as many people think. It's true that the first time you make it you are not going to know what to expect, but the next time you will. If you are blessed and the native yeasts on your apples are good fermenters, by paying attention to hygiene and following good techniques, you can expect something nearly the same the next season. The best tip, we think, is to keep these wild ferments cool. The cooler temperatures keep the fermentation moving slowly, giving the yeasts a chance to work their magic.

**YIELD:
3 GALLONS**

---

3 gallons (11.4 L) sweet cider

1 (12-ounce/350 mL) bottle dry cider

---

1. Sanitize a 3-gallon carboy and an airlock with a no-rinse sanitizer.

2. Measure the SG of the juice with a hydrometer and record it in your cider log or on a piece of masking tape attached to the carboy.

3. Add enough sweet cider to the carboy to fill within 3 to 4 inches of the top. Loosely cover the opening with a piece of plastic wrap.

4. Place the carboy on a surface that is easy to clean or set it on a tray or pan. Let it sit in an environment where the temperature is between 55°F/13°C and 65°F/18°C. Bubbles will slowly form after a few days, then build. If the liquid froths out the top and down the sides, simply clean the sides and tray/floor with a wet sponge. Wild yeasts are slow to get going so it could easily take a week or more until there are active bubbles. If it never froths over the side, that is okay, as long as the bubbles are active.

5. After the frothing, when the bubbles stay below the top of the carboy, apply the bung or lid, insert the airlock, and fill to the appropriate level with either fresh water or a neutral distilled spirit. The primary fermentation is finished when half to three-quarters of the sugars have been consumed, which you can determine by noting no bubbles being produced or by taking an SG reading. This will vary by yeast variety, but it generally takes 2 to 4 weeks. Note your initial SG reading to determine if it's reached the desired level. For example, if the initial SG was 1.060, now it would be 1.030 to 1.015 or lower. In our experience, this usually takes between 2 and 3 weeks, but can take up to a month or so. Taste the cider you used to measure the SG and write any tasting notes in your cider log or on a piece of tape attached to the carboy.

continued on next page

Wild Cider

continued

6. Using a syphon, rack your cider off into another sanitized 3-gallon carboy, making sure to draw off all the cider above the lees, without drawing the lees out.

7. Add enough bottled cider to top off the racked cider to within a couple of inches from the top of the new carboy to minimize air contact. Reapply the airlock.

8. Ferment in the same cool environment for 3 months.

9. After 3 months take one last SG measurement and calculate your final ABV. It will be rare that your SG will be more than 1.000. Taste, and if you would like it sweeter, back-sweeten it following the technique on page 106.

10. Siphon the cider into clean bottles, secure the tops, and store for at least 1 month before cracking one open. The bottles can be stored in a cool environment out of direct sunlight for a year or more, but at that point the cider will start to lose its sparkle, meaning it becomes a little thin tasting, losing the aromas and flavors and becoming something like carbonated apple water.

VARIATION: SCRUMPY

Scrumpy is not, as some have suggested, an inferior drink or a word for how it makes you feel after drinking it. The term comes from the United Kingdom and refers to a "rustic" cider, which usually means it's a bit cloudy and its apple lineage is difficult to pin down. It's made from a healthy mix of local apples. One of the definitions of *scrump* is "to steal apples from an orchard or garden," and we would assume that if you were filching apples, you would probably do it at night and that would make it pretty hard to see what you were picking, hence the mix. Press as many different types of apples as you can grab, then follow the Wild Cider recipe. Don't expect to duplicate the same scrumpy exactly next year, but that's the beauty of it.

MEET THE CIDERMAKERS
Bill Bleasdale & Chava Richman of Welsh Mountain Cider

Talking to Bill Bleasdale is like stepping inside the pages of his book *How to Grow Apples and Make Cider: Including Grow Pears and Make Perry* — lively, whimsical, and wide-ranging, with a dose of self-deprecation and a strong, yet simple, message throughout. Just do it.

Christopher came across his book through the recommendation of a cidermaker who said it's like no other cider book, and he was right. Bill didn't so much write it as draw it while he was accompanying his wife, Chava, who was working through the WWOOF program at Emandal farms in Northern California. Every illustration, every image, every word is hand-drawn. Bill is first and foremost an artist, one who, alongside Chava, happens to grow hundreds of varieties of fruit trees where most people wouldn't think they would thrive — in the East Cambrian mountains of Wales — and from them he makes some pretty amazing ciders.

Welsh Mountain Cider employs what they called a "traditional squeeze and leave" method of making their cider. They don't use sulfites or commercial yeasts. They don't rack either, preferring not to "muck about" but rather let the wild yeasts do

what they do best while leveraging their extensive knowledge of their apples and process. They finish their ciders to bone-dry, and some of their bottle-conditioned vintages are 4 to 6 years old and amazing. Bill's philosophy, which comes through strongly in both his book and in conversation, is not to be afraid of the process — or get too bogged down in the science of it. Nine out of ten times it will go right.

Bill told us that he believes taste is subjective and that people are awakening to the amazing complexity and flavors of natural ciders, which come from the apple varietals, their maturity, and orchard management. Welsh Mountain Cider and Tree Nursery has over 400 varieties of apples and 100 varieties of pears. After our conversation, we looked at all of our decisions, both in our orchard and in our cider house, with a new perspective that says: keep it simple, trust the natural process, and give beauty an equal share to efficiency or economics when making big decisions. Living a good life in the beauty of the natural world and keeping a good table are sometimes enough. In fact, they're everything.

Simple Sparkling Cider

As soon as you press fresh sweet cider, Mother Nature starts this recipe with the yeasts that came in and on the apples from the orchard. Even when stored in the refrigerator, fresh sweet cider will start to develop a little fizz after a week or more, which is an indicator that even in these less-than-optimal conditions, the yeasts are trying to do their jobs. In this recipe we break away from the 3-gallon glass carboys to reuse food-grade plastic bottles that are everywhere in our world. Our goal is to get enough carbonation and light alcohol development to have a sparkling, light, and sweet hardish cider.

YIELD: VARIES BY BOTTLE SIZE

Enough freshly pressed sweet cider to fill recycled plastic bottles

1. Wash and rinse your recycled plastic bottles, which can be anything from 1-gallon milk jugs to 16-ounce water bottles. Fill one bottle about one-third full with no-rinse sanitizer, close the cap, and shake for 20 seconds. You can pour the sanitizer into your next bottle and reuse three or four times. Invert the bottle to drain on a dish rack.

2. When the bottles are dry, fill them with cider to about 2 inches from the top and screw on the cap. Since the carbon dioxide has nowhere to escape, the bottle is going to start to expand. For that reason, we suggest you leave the bottle on your counter in plain sight, in a cool corner away from the heat of major appliances. Remember that wild yeasts like it on the cool side.

3. Depending upon the size of your bottles, the activity of the yeasts in your fresh juice, and the temperature, you should see the bottle swelling in a few days to a week or more. You are going to need to quickly "burp" this cider by partially unscrewing the cap, letting the gas escape, and quickly screwing it back on tight. You may end up doing this daily or twice a day, depending upon the amount of yeast activity and carbon dioxide being created. Let it sit in its cool corner for a week if you want it on the sweeter side, or up to a month if you want it dry.

4. Refrigerate this sparkling cider but know that it will continue to ferment even in the refrigerator, so you need to check it every few days and give it a wee burp.

MORE
POME
CIDERS AND
FRUIT
CIDERS

Apples aren't the only pome fruits that can be made into cider. Others are pears, some types of flowering roses, hawthorn, mayhaw, loquat, medlar, quince, and rowan berries — some of which grow on our farm and have yielded some interesting cider experiments.

According to the United States Association of Cider Makers (USACM), fruit ciders are "ciders with non-pome fruits or fruits/fruit juices added either before or after fermentation." There are many examples of fruit ciders out there today, including all types of berries, stone fruits (fruits with pits), and pineapples. The obvious reason to add these fruits is to bring out pleasing aromas and flavors in the cider, which can range from just a hint to right up front and center, pushing the apple right off the stage. They might also be added for additional color.

With the exception of perry, which is made from only pears, we make a fruit mash to mix in with the apple juice of each of the following fruit ferments. In our experiments, we preferred the widemouthed FerMonster carboy, especially when it came to cleaning, but you can use a bucket too. You will find other fruit ciders in the following chapter, all of which use cultured yeasts.

Perry

Perry is cider made exclusively from pears. The hardest part of this recipe is finding a few varieties of unsprayed pears. As with cider, you should make perry with what you have; sweet dessert pears are far more common in yards and gardens, but you will occasionally come across a large old tree with tiny bitter pears. It's likely one of the older varieties, planted by settlers or immigrants who were thinking of preserving the pears in perry, not of eating them fresh. Cidermakers with an interest in perry are also planting these tannic pears, because blending the flavor elements of the juices gives you a delicious complex perry. How do you know? Trust us: one bite of a cider pear and you will be able to discern tannins, a telltale sign of this distinctive pear.

You might be able to find pears in abandoned apple orchards. Long ago, at least in Oregon, many of the apple trees were grafted onto pear rootstock. The apple tree may be long gone, but the pear rootstock beneath will have turned into a tree often bearing little gnarled pears that make really good perry.

One last note: Part of what makes the taste of perry special is that pears contain sorbitol, which is an unfermentable sugar. For this reason, perry will retain a hint of sweet, even when finished completely dry. (For the less-sweet side of sorbitol, see page 151.)

YIELD: 3 GALLONS

---

3 gallons (11.4 L) freshly pressed pear juice

---

1. Sanitize a 3-gallon carboy and an airlock with a no-rinse sanitizer.

2. Measure the SG of the pear juice with a hydrometer and record it in your cider log or on a piece of masking tape attached to the carboy.

3. Add enough pear juice to the carboy to fill within 3 to 4 inches of the top. Cover the opening with plastic wrap. Pour the leftover juice in a bottle with a cap and keep in the refrigerator for use in topping off after racking. If you don't have at least 1 quart of leftover juice, you can use a quart of good quality organic preservative-free pear juice.

continued on next page

Perry
continued

4. Place the carboy on a surface that is easy to clean or set it on a tray or pan. Let it sit in an environment where the temperature is between 55°F/13°C and 65°F/18°C. Bubbles will slowly form after a few days, then build. If the liquid froths out the top and down the sides, simply clean the sides and tray/floor with a wet sponge. Wild yeasts are slow to get going so this could easily take a week or more until there are active bubbles. If it never froths over the side, that is okay, as long as the bubbles are active.

5. When the bubbles stay below the top of the carboy, apply the bung or lid, insert the airlock, and fill to the appropriate level with either fresh water or a neutral distilled spirit. The primary fermentation is finished when half to three-quarters of the sugars have been consumed, which you can determine either by noting no bubbles being produced or by taking an SG reading. This will vary by yeast variety, but it generally takes 2 to 4 weeks. Note your initial SG reading to determine if it's reached the desired level. For example, if the initial SG was 1.060, now it would be 1.030 to 1.015 or lower. Taste the perry you used to measure the SG and write any tasting notes in your cider log or on a piece of tape attached to the carboy.

6. Using a syphon, rack your perry off into another sanitized 3-gallon carboy, making sure to draw off all the cider above the lees, without drawing the lees out.

7. Add enough of the reserved pear juice to top off the racked cider to within a couple of inches from the top of the new carboy to minimize air contact. Reapply the airlock.

8. Ferment in the same cool environment for 2 months.

9. When ready to bottle, take one last SG measurement and calculate your final ABV. Taste, and if you would like it sweeter, back-sweeten it following the technique on page 106.

10. Siphon the perry into clean bottles, secure the tops, and store for at least 6 months. The bottles can be stored in a cool environment out of direct sunlight for a year or more. We have found that our perrys are better keepers than our apple ciders, and we have a couple of dusty bottles that are several years old and still delicious, not feeling the effects of age like the ciders from apples do.

A CALL FOR MODERATION

Sorbitol may lend a pleasant sweetness to perry, but the downside is that it will cause digestive upset if taken in large quantities. We heard a story from a very well-known and excellent winemaker who had a private contract to make perry (their first) for a client who supplied the cider pears. They made a lot of it — over a hundred cases — half barrel-aged in expensive new French oak barrels and half fermented in stainless steel and then infused with rose hips. The perry was on tap at a big party and guests came down with massive food poisoning symptoms soon afterward. The culprit wasn't actually food poisoning bacteria, but sorbitol. There is an old English saying that perry goes down slow and easy but comes out even faster (I'm sure you can guess what happened to those poor guests). Christopher has also experienced this unpleasantness when putting a nice perry in our 1-gallon carbonated growler and then drinking a couple of generous pints through the afternoon. So, you are forewarned: Go easy on the perry!

Quince Cider

We first met Eugene, a back-to-the-lander who'd come west in the early 1970s and lived up the road from us, when he dropped off a box of quince (*Cydonia oblonga*). It was also the first time we'd met quince. Eugene explained that he'd seen us and the kids working on the property and thought we might enjoy the quinces he'd picked at someone's house. Kirsten looked down into the cardboard box of sunny yellow fruit that were shaped like overweight, misshapen pears. The perfume was almost tropical, maybe like guavas. They were slightly fuzzy, like a peach, but the fuzz can be rubbed off easily as they ripen. She took them happily and started thumbing through old cookbooks to figure out what to do with them.

Quince are quite astringent and are not good for eating raw. Northern California wine country has pineapple quince orchards, which are now producing fruits for exciting quince ciders in those areas.

Many years after that first box of quince, we planted our own tree with thoughts of quince jam, but our small harvest every year becomes cider instead. Quince cider is robustly tropical and floral and can be made crystal clear with enough racking.

**YIELD:
3 GALLONS**

3¼ pounds (1.5 kg) quince

3 gallons (11.4 L) sweet cider

1. If your quince are fresh and firm, sweat them as you do apples (see page 59). They are done sweating when they feel waxy. They shouldn't show any browning. If they are already at this stage, proceed to the next step.

2. Sanitize a 3-gallon carboy and an airlock with a no-rinse sanitizer.

3. Quarter the quince, remove the stems and seeds, and finely chop in a food processor.

4. Add the chopped quince to the 3-gallon carboy and add the sweet cider, filling to the neck of the carboy. Swirl to mix. Reserve any remaining sweet cider in a 1-quart jar, secure with a lid, and refrigerate. We will use this in a few weeks when we rack the cider and need to fill some space in the carboy.

5. Measure the SG of the juice with a hydrometer and record it in your cider log or on a piece of masking tape attached to the carboy. It will probably be pretty close to the SG of the apple juice, because the quince won't be in the mood

continued on next page

yet to share anything. They will come around. Cover the opening loosely with a piece of plastic wrap.

6. Place the carboy on a surface that is easy to clean or set it on a tray or pan. Let it sit in an environment where the temperature is between 55°F/13°C and 65°F/18°C. Bubbles will slowly form after a few days, then build. If the liquid froths out the top and down the sides, simply clean the sides and tray/floor with a wet sponge. Wild yeasts are slow to get going so this could easily take a week or more until there are active bubbles. If it never froths over the side, that is okay, as long as the bubbles are active.

7. When the bubbles stay below the top of the carboy, apply the bung or lid, insert the airlock, and fill to the appropriate level with either fresh water or a neutral distilled spirit. The primary fermentation is finished when half to three-quarters of the sugars have been consumed, which you can determine by noting no bubbles being produced or by taking an SG reading. This will vary by yeast variety, but it generally takes 2 to 4 weeks. Note your initial SG reading to determine if it's reached the desired level. For example, if the initial SG was 1.060, now it would be 1.030 to 1.015 or lower. Taste the cider you used to measure the SG and write any tasting notes in your cider log or on a piece of tape attached to the carboy.

8. Using a siphon, rack your cider off into another sanitized 3-gallon carboy, making sure to draw off all the cider above the lees, without drawing the lees out. There will be a lot at the bottom.

9. Add enough of the reserved sweet cider to top off the racked cider to within a couple of inches from the top of the new carboy to minimize air contact. If you don't have enough juice, you can use commercial apple juice or, better yet, a bottle of your own homemade hard cider. Reapply the airlock.

10. Ferment in the same cool environment for at least 2 months, though we have found that this cider does well with a long secondary ferment of a full 3 months.

11. Rack your cider again, following the instructions in step 8. Let mature for an additional 1 to 2 months.

12. Take one last SG measurement, which should be at 1.000 or below, and calculate your final ABV. Taste, and if you would like it sweeter, back-sweeten it following the technique on page 106.

13. Siphon the cider into clean bottles, secure the tops, and store for at least 1 month (hey, what's one more month after all this fermentation time?). The bottles can be stored in a cool environment out of direct sunlight for a year or more, but at that point the cider will lose some of its sparkle.

Medlar Ginger Cider

When we first moved onto our land there was a small local nursery that specialized in rare and unusual plants. We bought things we had never heard of, many of which didn't live, but the medlar (*Mespilus germanica*) did. We were intrigued by its history as a medieval fruit that had fallen out of favor due to, among other things, the fact that it can't be eaten until it's nearly rotten. And how can one not be intrigued by a fruit that has been so unflatteringly portrayed by writers like Shakespeare, Chaucer, and D.H. Lawrence?

The flavors in this cider are reminiscent of gingerbread. The heat of the ginger comes through nicely, and the wild yeasts keep just a tiny amount of residual sugar in place. If you cannot procure medlars, you can also make a pure ginger cider, which we have done often, and it is simply delicious. Conversely, you can make a pure medlar cider — just omit the ginger and you will get a cider a little like a perry. Either way, the instructions are the same. As you may suspect, a mushy medlar will not make juice; you'll get a paste.

**YIELD:
3 GALLONS**

8 pounds (3.65 kg) bletted medlar fruit (see page 157)

1 cup (237ml) ginger juice (from about 11 ounces/312 g fresh ginger)

3 gallons (11.4 L) unpasteurized sweet cider

1. Sort through the bletted medlars (see page 157) and discard any rotten or moldy fruit. Rinse them with water and place them in a very large bowl. Using a potato masher, smash the medlars into a paste.

2. Sanitize a 3-gallon carboy and an airlock with a no-rinse sanitizer.

3. Add the medlar mash and ginger juice to the carboy. Add enough sweet cider to the carboy to fill within 3 or 4 inches of the top. Swirl to mix. Reserve any remaining sweet cider in a 1-quart jar, secure with a lid, and refrigerate. We will use this in a few weeks, when we rack the cider and need to fill some space in the carboy.

4. Measure the SG of the juice with a hydrometer and record it in your cider log or on a piece of masking tape attached to the carboy. It will probably be pretty close to the SG of the sweet cider, as the medlars won't be in the mood yet to share anything. They will come around. Cover the opening with plastic wrap.

continued on next page

5. Place the carboy on a surface that is easy to clean or set it on a tray or pan. Let it sit in an environment where the temperature is between 55°F/13°C and 65°F/18°C. Bubbles will slowly form after a few days, then build. If the liquid froths out the top and down the sides, simply clean the sides and tray/floor with a wet sponge. Wild yeasts are slow to get going so this could easily take a week or more until there are active bubbles. If it never froths over the side, that is okay, as long as the bubbles are active.

6. When the bubbles stay below the top of carboy, apply the bung or lid, insert the airlock, and fill to the appropriate level with either fresh water or a neutral distilled spirit. The primary fermentation is finished when half to three-quarters of the sugars have been consumed, which you can determine either by noting no bubbles being produced or by taking an SG reading. This will vary by yeast variety, but it generally takes 2 to 4 weeks. Note your initial SG reading to determine if it's reached the desired level. For example, if the initial SG was 1.060, now it would be 1.030 to 1.015 or lower. Taste the cider you used to measure the SG and write any tasting notes in your cider log or on a piece of tape attached to the carboy.

7. Using a siphon, rack your cider off into another sanitized 3-gallon carboy, making sure to draw off all the cider above the lees, without drawing the lees out.

8. Add enough of the reserved sweet cider to top off the racked cider to within a couple of inches from the top of the new carboy to minimize air contact. If you don't have enough sweet cider, you can use commercial apple juice or, better yet, a bottle of your homemade hard cider. Insert the airlock and fill it to the appropriate level with either fresh water or a neutral distilled spirit.

9. Ferment in the same cool environment for at least 3 months or up to 6 months. We have found that this cider does well with a long secondary ferment.

10. If you want a crystal-clear cider, rack it again following the instructions in step 7. Or use the *sur lie* (page 102) or the bâtonnage (page 104) method of aging. Let mature for an additional 1 to 2 months.

11. Take one last SG measurement, which should be at 1.000 or below, and calculate your final ABV. Taste, and if you would like it sweeter, back-sweeten it following the technique on page 106.

12. Siphon the cider into clean bottles, secure the tops, and store for at least 1 month (hey, what's one more month after all this fermentation time?). The bottles can be stored in a cool environment out of direct sunlight for a year or more, but at that point the cider will lose some of its sparkle. The flavor will likely change over the year, so plan to taste one every month and find your happy place, then note that in your cider log for next year's batch.

MEET THE MEDLAR

There is nothing about this fruit, which originated in Persia and has likely been cultivated for three thousand years, that works with our modern food systems. Our now 20-year-old medlar is a small tree that grows more diagonally than upright, more likely due to our poor location choice than the habit of the tree. It is quite prolific. It blooms a month later than the apple trees and has beautiful white flowers. The small fruits are a golden rusty green to a rosy brown and look a bit like a cross between a round rose hip and a small apple. In fall, as the leaves turn red and drop to the ground, these little orbs hold on, decorating the bare branches.

Once the tree started producing fruit, it was another few years before we figured out how to eat them. We'd read that, like persimmons, cornelian cherries, quince, and other astringent fruits, they must be "bletted," meaning they should be allowed to soften past the ripe stage. Medlars are usually picked right after the first frost and then left in a cool, dark place for several weeks until they are at the edge of decay. At this point, the flesh is brown (not quite overripe banana but definitely not crispy) and can be spooned out of the skin, while avoiding the seeds. The bletting increases the sugars and decreases the acid and astringent tannins.

While we enjoyed the custardy flavor of our bletted medlars, they never seemed quite right. Some were too dry by the time they were soft. Then one year we left the fruit on the tree. In January, we started eating them right off the tree as we walked past them to do chores. We picked as many as we could carry and squeezed the sweet, soft, and cold puddinglike fruit, that had been frozen and thawed many times, out of the stem end into our mouths, dropping skins and seeds along the way. These fruits also make a tasty cider.

Cornelian Cherry Cider

We love this cider. It's clear and rosy and has a beautiful, bright flavor. And it solved a problem for us. When we planted a food forest a number of years ago on a terrace behind our home, cornelian cherries (*Cornus mas*) filled a lower-story niche in the chestnut guild. We ordered the 'Elegant' and 'Pioneer' cultivars and they have done very well — so well, in fact, that we haven't known what to do with all the tasty, but very astringent, fruit. When made into cider, the astringency and tartness are a plus. If this cider is too tart for you, allow it to go through a malolactic fermentation before bottling. Should you never come across cornelian cherries, don't despair. You can make this cider with any sour pie cherry, as pictured.

**YIELD:
3 GALLONS**

---

2 quarts (1.1 kg) fresh or frozen cornelian cherries, or other sour cherries

3 gallons (11.4 L) unpasteurized sweet cider

---

1. Sanitize a 3-gallon carboy and an airlock with a no-rinse sanitizer.

2. In a very large bowl and using a potato masher, coarsely mash the cherries without damaging the pits.

3. Add the cherry mash (cherry flesh and pits) to the carboy. Add enough sweet cider to the carboy to fill within 3 or 4 inches of the top. Swirl to mix. Reserve any remaining sweet cider in a 1-quart jar, secure with a lid, and refrigerate. We will use this in a few weeks, when we rack the cider and need to fill some space in the carboy.

4. Measure the SG of the juice with a hydrometer and record it in your cider log or on a piece of masking tape attached to the carboy. Cover the opening with plastic wrap.

5. Place the carboy on a surface that is easy to clean or set it on a tray or pan. Let it sit in an environment where the temperature is between 55°F/13°C and 65°F/18°C. Bubbles will slowly form after a few days, then build. If the liquid froths out the top and down the sides, simply clean the sides and tray/floor with a wet sponge. Wild yeasts are slow to get going so this could easily take a week or more until there are active bubbles. If it never froths over the side, that is okay, as long as the bubbles are active.

continued on next page

Cornelian Cherry Cider

continued

6. When the bubbles stay below the top of the carboy, apply the bung or lid, insert the airlock, and fill to the appropriate level with either fresh water or a neutral distilled spirit. The primary fermentation is finished when half to three-quarters of the sugars have been consumed, which you can determine by noting no bubbles being produced or by taking an SG reading. This will vary by yeast variety, but it generally takes 2 to 4 weeks. Note your initial SG reading to determine if it's reached the desire level. For example, if the initial SG was 1.060, now it would be 1.030 to 1.015 or lower. Taste the cider you used to measure the SG and write any tasting notes in your cider log or on a piece of tape attached to the carboy.

7. Using a siphon, rack your cider off into another sanitized 3-gallon carboy, making sure to draw off all the cider above the lees, without drawing the lees out.

8. Add enough of the reserved sweet cider to top off the racked cider to within a couple of inches from the top of the new carboy to minimize air contact. If you don't have enough sweet cider, you can use commercial apple juice or, better yet, a bottle of your homemade hard cider. Reapply the airlock.

9. Ferment in the same cool environment for 2 months.

10. Take one last SG measurement and calculate your final ABV. Taste, and if you would like it sweeter, back-sweeten it following the technique on page 106.

11. Siphon the cider into clean bottles, secure the tops, and store for at least 1 month before cracking one open. The bottles can be stored in a cool environment out of direct sunlight for a year or more, but at that point the cider will lose some of its sparkle.

American Persimmon Cider

This clear, semisweet cider is pure persimmon in a bottle. When we drink this in summer, we are transported to golden, late-fall days when these sunset orange orbs decorate the branches. We make this cider with tiny American persimmons, which we planted as seedlings a number of years ago. While they don't grow wild in the Pacific Northwest, the fruits are native to America and can be foraged in other parts of the country. You could also substitute Texas persimmons or Hachiya, the astringent Japanese persimmon. Allow the persimmons to soften before pressing.

Our recipe varies depending on our persimmon harvest, but our favorite is to use a ratio of 1 pound of persimmons to 1 quart of apple juice. Our persimmons often come in later in fall, when our apples have already been squeezed into juice and placed in carboys, so we often infuse them into our cider at the second fermentation. When adding them to the primary fermentation, as in this recipe, we prefer to use pasteurized apple juice because the wild yeast on the persimmons has a lower alcohol tolerance and leaves the cider with a small amount of residual sweetness. You can also use freshly pressed sweet cider; the cider will just finish a little dryer.

**YIELD:
3 GALLONS**

---

5 pounds (2.27 kg) softened American persimmons

2½ gallons (9.5 L) pasteurized apple juice

---

1. Sanitize a 3-gallon carboy and an airlock with a no-rinse sanitizer.

2. Remove the stems and leaves from the persimmons. In a very large bowl, using a potato masher, coarsely mash the persimmons.

3. Add the persimmon mash to the carboy. Add enough apple juice to the carboy to fill within 3 to 4 inches of the top. Swirl to mix. Reserve any remaining apple juice in a 1-quart jar, secure with a lid, and refrigerate. We will use this in a few weeks, when we rack the cider and need to fill some space in the carboy.

4. Measure the SG of the juice with a hydrometer and record it in your cider log or on a piece of masking tape attached to the carboy. Cover the opening with plastic wrap.

continued on page 163

5. Place the carboy on a surface that is easy to clean or set it on a tray or pan. Let it sit in an environment where the temperature is between 55°F/13°C and 65°F/18°C. Bubbles will slowly form after a few days, then build. If the liquid froths out the top and down the sides, simply clean the sides and tray/floor with a wet sponge. Wild yeasts are slow to get going so this could easily take a week or more until there are active bubbles. If it never froths over the side, that is okay, as long as the bubbles are active.

6. When the bubbles stay below the top of carboy, apply the bung or lid, insert the airlock, and fill to the appropriate level with either fresh water or a neutral distilled spirit. The primary fermentation is finished when half to three-quarters of the sugars have been consumed, which you can determine by noting no bubbles being produced or by taking an SG reading. This will vary by yeast variety, but it generally takes 2 to 4 weeks. Note your initial SG reading to determine if it's reached the desired level. For example, if the initial SG was 1.060, now it would be 1.030 to 1.015 or lower. Taste the cider used to measure the SG and write any tasting notes in your cider log or on a piece of tape attached to the carboy.

7. Using a siphon, rack your cider off into another sanitized 3-gallon carboy, making sure to draw off all the cider above the lees, without drawing the lees out.

8. Add enough of the reserved apple juice to top off the racked cider to within a couple of inches from the top of the new carboy to minimize air contact. Insert the airlock and fill it to the appropriate level with either fresh water or a neutral distilled spirit.

9. Ferment in the same cool environment for 2 months.

10. Take one last SG measurement and calculate your final ABV. Taste, and if you would like it sweeter, back-sweeten it following the technique on page 106.

11. Siphon the cider into clean bottles, secure the tops, and store for at least 3 months before cracking one open. The bottles can be stored in a cool environment out of direct sunlight for a year or more. We have found this cider ages quite well.

SOUR
CIDERS

Sour ciders are gaining appeal. They break a cidermaking rule of embracing only yeasts and rejecting bacteria. The bacteria in question — lactic acid bacteria (LAB) — are better known for producing the pickle-y flavor in fermented vegetables. They, as well as naturally acidic apples, are responsible for the sour taste in cider. We find it interesting that you use roughly the same process to make a sour cider as you do to make a Belgium-based lambic beer. We think of this style of cider as a bridge between sweet cider and where cider wants to go naturally — vinegar — but in a very enjoyable, drinkable way.

Spanish-style cider, called sidra, is one example of a sour cider. It embraces the microbes that other styles shun — specifically, lactic and acetic acid bacteria. Sidra is sparkling, very fruity, and usually has a lower alcohol level. It is pale to a light gold in color. Traditional sidra is made by picking the apples late in the season and off the ground, a practice that is still allowed in Europe but not in the United States. The apples are pressed, and the fresh juice is fermented in large open tanks, where it is exposed to oxygen and the vinegar bacteria, giving the cider acetic aromas and flavors. The leading cidermakers we visited in northern Spain were as modern as many in the United States, though some of the aging barrels are much, much older and bigger. In fact, we were a bit surprised by what we found. Expecting to find traditional open wooden tanks, we toured cideries with modern stainless steel with sanitation protocols. For these Spanish cidermakers, the sour comes from a careful balance of very sharp apples, malolactic fermentation, and a third fermentation phase — not from sustained exposure to air from open wooden tanks to encourage the development of acidic acid. In fact, the system we saw, developed by Miguel Zapiain, allows for the juice to be in contact with air for even less time than the average cidery.

Traditional sidra is finished to full dryness and is still. It goes through a slow, wild yeast fermentation process, then it is matured in chestnut kegs to 4.5 to 6.5 percent ABV before being bottled with no added carbonation. A little carbonation in the glass comes from the height from which the cider is traditionally poured from the bottle or keg, depending upon the region. Some of the famous flavors come from the seasonality of fermentation and temperature. These unpasteurized ciders were traditionally sold in January, after the buyers chose the ones they wanted bottled, straight from the barrels. The unsold ciders were consumed through about April, at which point they were bottled for the rest of the season. By the following fall they'd often lost a lot of quality because they continued to ferment in the bottle. They were then sold very cheaply during apple harvest and cidermaking festivals to get rid of them before the next batch. Sidras are tart, fun, and range from mildly to aggressively funky, though we had some in both regions with no funk at all.

Miguel & Ion Zapiain of Zapiain Sagardotegia

The cider operation at Zapiain, in the heart of the Basque cider country at Astigarraga (Gipuzkoa) in northern Spain, was closed to the public when we visited in the latter part of May. This region strengthened as a cider region because of its proximity to the historic port in San Sebastián (Donostia), where barrels were loaded up and floated down the river to load onto the waiting ships — drink for the sailors. Traditionally, Basque cider houses (*sagardotegi* in the Basque language) are only open to the public from January through April, when this style of cider, known as sagardoa, is at its best. For many centuries, people from the surrounding communities would bring food to share, as well as enjoy food provided by the cider houses, when everyone came together to sample the year's offerings. The word for this meal and communal tasting is *txotx*, which is pronounced "choach," and we were pretty sad we missed it. Ion Zapiain and his father, Miguel, made up for it, however, by leading us through an amazing behind-the-scenes tour of the facility, including tasting many of their ciders in various states of maturity in massive stainless steel tanks.

The Zapiain family have been making cider for a very long time. They list 1595 on their label because that was the year that a direct ancestor, Juanes de Zapiain, was officially fined and told not to bring his cider to town because he sold cider in the neighboring port town of San Sebastián (Donostia) without paying the appropriate taxes. Miguel's father built the current facility in 1961, a time when cider consumption per capita was at historic lows and it looked as if cider might fade from the culture altogether. During this time the Zapiain capacity nearly matched the total cider consumption of the region. Miguel's father had a vision that Miguel would build upon. Miguel was the first person in the family to receive formal education in the science of wine, and after studying abroad, he returned to the family business as an oenologist. He has devoted the last

35 years to not only making Zapiain's product the best quality but also raising the quality of all Basque Country cider. That has meant continuing education of fellow cidermakers and also of consumers. Consumers just like us.

We traveled to Spain with two goals: to teach fermentation (including cidermaking) and to understand Spanish-style cider. We thought we knew what the style was, and we honestly love the funkiness that we associate with this type of cider. Drinking our way through the Basque and their Asturias regions, we were not surprised nor disappointed, though only one of the bottles was majorly funky, reminding us of a lovely blue cheese. Then we met these master cidermakers who are devoted to changing that definition for the world. After tasting our way through their fermentation rooms, we understood why. Their cider was bright and lightly sour with lingering apple notes. As Miguel told us through Ion's translation, "We must work properly with the apple," he said. "Everything we need is in the apple."

They have invented new equipment and processes that maximize the hygiene of the operation — from more efficient transportation of apples and multiple washing stations to an ingenious and elaborate system of progressively larger reservoirs that minimizes the contact the cider has with the air immediately after pressing. Their philosophy is simple: no preservatives, no pasteurization, and very high-quality standards. Miguel pays close attention to each stage of the fermentation.

The results are impressive. The cider is clean and fruit-forward, with a high acidity yet no funk, no barnyard — no off anything. And at just 6 percent ABV and traditionally served at less than 4 ounces per glass, it's something that you could see yourself enjoying throughout a long summer day. Or inside in wintertime at the cidery, straight from the massive wooden barrels, accompanied by traditional cod omelets, fried cod, steak, cheeses, and walnuts. That is, if you visit in the right month.

Spanish-Style Cider

There are two types of cider produced in Spain: a sparkling cider and a natural cider. While the sparkling allows for concentrated juices, added sugars, and forced carbonation, the natural cider is the one we are making here. Made according to traditional methods, the natural ciders have no added sugars or carbonation and are made only from pressed cider apples. These apples are typically higher in tannins and more acidic than apples for fresh eating.

The most unique characteristic of Spanish-style ciders is their slightly acetic character, which most believe comes from the same acetobacter bacteria that make vinegar. The trick is to get only a touch of acetic or lactic acid because otherwise, it's a one-way road to funk and vinegar. This can be accomplished is by constantly adding new juice for a few days, thereby aerating the cider that has begun to ferment, or in this recipe, by stirring. This gives the acetic bacteria a little start before they are turned off by the full cask and anaerobic fermentation (anaerobic being the key!). The other, perhaps "safer" way that we learned about in Spain, is to use only tart and bitter apples.

If you are pressing the juice yourself, allow the mash to macerate in buckets for 8 to 24 hours before pressing.

**YIELD:
3 GALLONS**

---

3 gallons (11.4 L) freshly pressed sweet cider from predominately tart apples (Granny Smith, McIntosh, Jonathan, or sharp cider apples if you can get them)

1 (12-ounce/350 mL) bottle dry cider

---

1. Sanitize a 3-gallon food-grade plastic bucket with a no-rinse sanitizer.

2. Measure the SG of the sweet cider with a hydrometer and record it in your cider log or on a piece of masking tape attached to the bucket. Measure the pH with a pH test strip and record it in your cider log or on the masking tape.

3. Pour the sweet cider into the sanitized bucket and cover loosely with a lid. Let it sit for 12 hours in a cool environment that is between 55°F/13°C and 65°F/18°C. The acetobacters like a warmer environment so we want to keep this cider as cool as possible within this temperature range to favor the wild yeasts.

4. Gently stir with a sanitized long spoon for 30 seconds. Replace the lid and let it sit for another 2½ days.

5. Sanitize a 3-gallon carboy and an airlock with a no-rinse sanitizer.

6. Pour the sweet cider from the bucket into the sanitized carboy to fill within about 2 inches of the top. Insert the airlock and fill it to the appropriate level with either fresh water or a neutral distilled spirit.

7. When you no longer see bubbles in the airlock, the primary fermentation cycle is likely done. This will vary by yeast variety, but it generally takes 2 to 4 weeks. Take another specific gravity reading. Note your initial SG reading to determine if it's reached the desired level. For example, if the initial SG was 1.060, now it would be 1.030 to 1.015 or lower. Taste the cider you used to measure the SG and write any tasting notes in your cider log or on a piece of tape attached to the carboy. Taste, and see if you can already pick up on the sour notes.

8. Using a siphon, rack your cider off into another sanitized 3-gallon carboy, making sure to draw off all the cider above the lees, without drawing the lees out.

9. Add enough bottled cider to top off the racked cider to within a couple of inches from the top of the new carboy to minimize air contact. Insert the airlock and fill it to the appropriate level with either fresh water or a neutral distilled spirit.

10. Ferment in the same cool environment for 3 months or up to 6 months to allow plenty of time for the malolactic fermentation to fully take place, which will mellow out the high acidity of this cider.

11. Take one last SG measurement and calculate your final ABV. Taste some of the cider used to measure SG. You should pick up some sour notes but not so much that you think of vinegar.

12. Siphon the cider into clean bottles, secure the tops, and store for at least 1 month.

THE HEIGHT OF THE POUR

In northern Spain, naturally still cider is carbonated briefly by the unique and beautiful technique for serving it. The method depends on the region. In the Asturias region of northwestern Spain, where 80 percent of the country's cider is produced, an *escanciador* performs the pour (*escanciar un culin*) by holding the cider bottle above his head and pouring while staring straight ahead, so that an arching stream falls into a small widemouthed drinking glass, held as low as possible in the other hand. This moment of theater is beautiful to watch, as the *escanciador* closes his eyes to center himself before beginning the pour. It is a tad messy but it adds a bit of aeration that gives your mouth the feeling of carbonation, though the feeling fades quickly, so you are encouraged to drink it quickly in one gulp. The tradition of sharing a glass among those at the table has sadly faded due to health restrictions, so expect to have your own glass.

In the Basque Country, the cider is traditionally served directly from very large aging barrels. A small hole is plugged with putty and opened with what could be described as a modified ice pick. When this is pulled, a thin, arching stream of cider gushes forth, which is caught by a line of eager drinkers waiting with their glasses to scoop up about a quarter glass of cider before the person behind them continues catching the stream. Both styles encourage anticipation, moderation, and community.

KEEVING

Making keeved cider is an experience and tasting it (when it works) is sublime. Keeving is a traditional French way of fermenting apple juice that leaves the residual sugars of the apple, producing a semi-dry sparkling cider — using just apple juice and a little calcium chloride a few days in. It's an art. A keeved cider is about slowing down. It will take plenty of your time and attention for about a year. The flavors are often pure and complex —tannic, naturally sweet, and a little bit funky. In short, it's a wonderful expression of the apple. What's not to love about that? Well, it's complicated.

Chapeau brun *formed on a bucket of keeving cider*

The keeving process is dependent on the formation of an abundant brown gelatinous crust on top of the juice. This cap, called the *chapeau brun*, needs to form before the yeasts wake up and fermentation begins. This is important because this little brown hat traps most of the impurities, a good bit of nitrogenous substances, and many yeast cells. This creates an environment in the fermenting must below that has fewer yeasts and yeast nutrients, slowing the fermentation way down. Commercial yeasts or cultured nutrients are never added. The hope is that the fermentation eventually becomes "stuck." It is this inability for the yeasts to continue that makes a stable cider that retains its sweetness. It's as simple as that.

INSURANCE POLICY: WILD YEAST STARTER

It is important to *never* add commercial yeast to keeved cider. *Never*. We usually have a looser attitude toward recipes and processes, believing it's important to allow everyone's creativity full freedom, but in this case, we mean it. This a wild yeast party and the commercial yeasts are not invited because they just won't halt as needed. It's a delicate balance — the ferment must be held back long enough to create the cap but be gentle enough to stop with residual sugars, and vibrant enough to do its job. There might be a point when the fermentation is too slow. If you premake a yeast starter (page 211) with the same pressed apples before you start the keeve batch, you will have some yeast to pitch should things get stuck.

The *chapeau brun* is formed when calcium is added to the must. Calcium was traditionally added in the form of chalk or ashes, to which was added some table salt. If you cannot find food-grade calcium chloride ($CaCl_2$), available through cheese-making suppliers and some brew shops, you can use calcium carbonate ($CaCo_3$), available in the supplement section of natural foods stores, and table salt. The pectin methylesterase (PME) enzymes in the apple flesh transform pectinic acids in such a way that when calcium is added, it creates an insoluble gel at the bottom of the barrel. Those of you who make homemade jams and jellies know how important this jelling quality is. The small bubbles of carbon dioxide from the beginning of the slow fermentation get trapped in the gel and cause it to start migrating up to the surface. These bits compact at the top, along with any hitchhikers, such as the above-mentioned impurities or stray yeasts, and violà — *chapeau brun*. It is important to note that you will lose a good portion, about 20 percent, of the juice to this cap.

Back in the day, imagine cidermakers operating giant screw presses consisting of a stone wheel pulled by a beast, piles of apples on the ground — apples grown for their ability to keeve. The apples we grow now don't consistently give enough PME, so that must be added to the juice. (Fun fact for those of you familiar with our last book, *Miso, Tempeh, Natto & Other Tasty Ferments*: PME is made from a fungal relative of *Aspergillus oryzae* (koji): *A. niger*.) There were times in our initial testing where we could not obtain PME (see Suppliers, page 321). We tried running some pressed mash through our Champion juicer and adding that to the regularly pressed juice to up the pectin, but only a thin cap formed; we recommend adding PME.

Some cidermakers add sulfites to slow the beginning of the fermentation, but we have never done this, and it is not traditionally done in France. But if the temperatures are higher than optimal, or if you are unable to bring a low-acid cider (3.7 to 4.0) to an acceptable acidity, you may wish to consider this. For those who would like to try it, add half a normal dose of sulfite to kill off part of the wild yeast population to give time for the *chapeau brun* to rise. The risk,

GENERAL GUIDELINES
FOR ADDING SULFITES TO A KEEVE

If the juice has a pH of 3.7 to 4.0, crush and dilute 1½ Campden tablets (potassium metabisulfite) per 1 gallon of juice. If the juice has an optimal pH (3.4 to 3.6), you can use up to ¾ tablet for 1 gallon of juice, but remember there is a lot of risk associated with this method. If the acid is high and the pH reads 3.3 or below, the juice is too acidic.

of course, is that you will kill off too many of the yeasts, making a perfect environment for mold to have a party and kill the keeve. (If mold is discovered soon after forming, it is possible to salvage the keeve by removing the mold and adding commercial yeast; otherwise you'll need to dump it and start again.)

The formation of the cap is based on the quality of the pectin and the temperature. You can lower the temperature to impede the fermentation if needed or raise it to increase fermentation slightly. Remember: easy does it — just a degree or two at a time. If the fermentation is too active it will destroy the cap before it rises. You can set a radiant space heater at a very low temperature next to your bucket for a short period of time. Monitor it constantly so that you can remove it when the temperature rises. Place an inexpensive thermometer strip on your vessel, so you can have a sense of the temperature without prying into the contents of the container. If fermentation still doesn't get under way, avoid the temptation to pitch a commercial yeast, even a small amount. This will cause a fermentation that can't be halted as needed, which is part of why we keeve. If you have made a wild yeast jar, page 211, you can use it here.

Once the gel has buoyancy it will rise, which can happen in many pieces or as one whole piece. If, after a couple of weeks, no *chapeau brun* forms, it failed to

Don't Get Sharp with Me

CHRISTOPHER WRITES: I haven't spent a lot of time in a lab and that's probably why I am still here and whole. Something about all those chemicals, burners, sharp objects, and glass makes me want to yell FIRE! and jump out a window. So when we had mandatory lab day in our cidermaking class, I was only looking forward to wearing the white lab coat. Our simple task was to measure our juice blend's pH and titrated acidity. My blend wasn't actually a blend because I was talking during the time we were supposed to blend, so by the time I got to the table there was only a gallon of Redstreak juice. Being lazy, I decided that would be my blend, but when I measured the acidity of this notoriously sharp apple it was really acidic — so much so that I was going to need to dump in a lot of calcium carbonate to balance it. When our instructor saw me going for all of the calcium carbonate, he intervened to test my numbers, but they proved out. It turns out there are no medals for the sourest cider in the world.

keeve, but not all is lost: you can rack the juice into a carboy and pitch it with a commercial yeast and still save the cider. It won't be a keeved cider, but it will likely be a good cider. If a *chapeau brun* forms, congratulate yourself and keep reading.

This cider ferments a bit differently from other ciders. Instead of a squall in a bottle of wildly active primary fermentation, this will look more like the gentle movement of a secondary ferment. If it is too active you will want to slow it down. Keep it cool — between 33°F/1°C and 41°F/5°C. We can't say this enough.

Racking is used frequently in keeving, at optimal moments, to stabilize the ferment and produce the desired flavors. This is about bubble watching and slowing down the ferment. Each time you rack, you will eliminate more yeast and help to stabilize the ferment. As you are racking, check the SG. Cool off the juice and rack more if the fermentation is proceeding too quickly — temperature and racking are your best natural tools. Plan on racking about three times and make sure you top off the carboy each time using reserved cider saved from the previous racking. (You can keep this fermenting in a smaller bottle.)

SIMPLE PECTIN TEST

This simple test can be done fairly quickly and gives you a sense of how much soluble pectin is in your juice. The test is from Claude Jolicoeur's book *The New Cider Maker's Handbook*, in which Claude notes that the test is for making jelly and the results aren't easily interpreted. We take that to say a successful test does not guarantee a successful keeve, but it may help you determine if you have enough soluble pectic acids for a snowball's chance in a warm spot of forming a *chapeau brun*. And it is fun kitchen chemistry.

Take a sample of strained juice and mix it with two or three times the volume of strong alcohol (rubbing alcohol or Everclear). After a few minutes it should form a gel that will be more or less solid (a little brown cap); this is a good sign. If the pectic acid content is low, only strands of pectin will form.

Keeved Cider

Make this cider your last batch of the season. Here's why: you want to use late-season apples, as these apples are often (though not always) a little slower to start fermenting. If you happen to have a choice in apples, then look for varieties that are known for having low nutrients (nitrogen, especially) and high pectins. Here's a hint: apples from unfertilized, stressed older trees will have lower nitrogen. Late-ripening bittersweets, bittersharps, and sweetsharps have high SG and high pectin. The apples should be beyond ripe — in other words, soft and spongy in texture, but not brown and rotting.

You also need cold weather. As you are pressing and starting the ferment, it's important that the temperature be cold because you really don't want to wake up the yeasts until the *chapeau brun* has formed. Ideally, the temperature in the space where you are pressing should be no higher than 50°F/10°C. When your apples are ready and the conditions are favorable, plan on spending a few days minding the ferment.

Here we've given a set of instructions and a guideline to the process more than an actual set recipe; adjust the number of apples as needed to fill your containers. This assumes you'll have about 4 gallons of juice in a 5-gallon bucket to be put in a 3-gallon carboy. (Remember you lose juice to the *chapeau brun* and lees.)

**YIELD:
2–3 GALLONS**

---

80 pounds (36 kg) apples, or thereabouts, sweated, washed, and sorted (see note)

¼ teaspoon (2 g) pectin methylesterase (PME) diluted in 1 cup (237 ml) apple juice

5½ teaspoons (28 mL) food-grade calcium chloride liquid solution

---

1. Sanitize two 5-gallon food-grade plastic buckets with a no-rinse sanitizer.

2. Grind up the apples (see page 60) to create a mash.

3. Divide the mash between the sanitized food-grade plastic buckets and let it sit for at least 3 and up to 24 hours in a cool space, ideally 33°F/1°C to 41°F/5°C and no warmer than 50°F/10°C, such as the refrigerator. The cooler the temperature, the longer you can let it sit. This step slows down oxidation and helps to free the pectin from the skins in the mash.

4. Press the mash into sweet cider (see page 63). If desired, use a simple pectin test to check the cider's keeving ability (see box on page 173).

5. Sanitize another 5-gallon food-grade plastic bucket with a no-rinse sanitizer. Pour the sweet cider into the bucket, leaving about 4 to 5 inches of space at the top of the bucket, where the *chapeau brun* will form. Add the diluted PME

to the sweet cider and stir. Reserve any remaining sweet cider in a 1-quart jar, secure with a lid, and refrigerate for topping off after racking.

6. Measure the SG, and the pH if desired, and record it in your cider log or on a piece of masking tape attached to the bucket. The SG should be at 1.055 or above. If lower, add cane sugar until you reach an acceptable level. For every 0.005 increase, you will need to add 2.25 ounces of sugar for every gallon of juice. (See How to Sweeten Your Juice, page 71.) The pH should be between 3.4 and 4.0 (see Note on page 68 for testing methods). If the pH is too high, blend in a more acidic apple juice; if that is not possible, add malic acid, which is the best choice, or citric acid. If the pH is too low, blend in some nonacidic apple juice, nonchlorinated water, or calcium carbonate, also called precipitated (or ground) chalk. Cover the bucket with the lid, which should be tightfitting but not fully sealed.

7. Let the cider sit for 1 to 2 days in a cool environment: between 33°F/1°C and 41°F/5°C. Stir once or twice to distribute the PME.

8. Add the calcium chloride solution and stir for several minutes.

9. Ferment, undisturbed, in the same cool environment for 7 to 18 days as the *chapeau brun* forms. Agitation can cause the *chapeau brun* to drop, so try not to move it around. It is also helpful to ferment in a spot where you can adjust the temperature as needed. Check it daily. (If you see mold forming, rack it immediately and introduce a commercial yeast; you will no longer be keeving cider, but you will save your juice.)

10. Check the SG. A medium-sweet cider would stabilize at 1.018.

11. Poke through the edge of the cap with a sanitized wine thief or a turkey baster to make sure the must is clear with no suspended gel. Using a siphon, rack it into a sanitized 3-gallon carboy with an airlock. If needed, top off the bucket with the reserved sweet cider.

NOTE: Remove any apples that are blemished by worm holes, bruises, or punctures. You must be a little more persnickety than with a regular wild cider because it will take some time before the fermentation process can kill off any "bad" microbes.

12. Age for 2 to 3 months for medium-sweet cider, or longer for dry cider. Rack twice during this time to reduce the yeast.

13. Take one last SG measurement, which should be at 1.015 for medium-sweet cider, and calculate your final ABV.

14. Siphon the cider into clean bottles, secure the tops, and store for at least 1 month. The bottles can be stored in a cool environment out of direct sunlight for 1 year or more, but at that point the cider will lose some of its sparkle.

CHAPTER 5

CULTIVATED CIDERS

Deciding to use a commercial yeast doesn't mean you need to carpet bomb all the natural yeasts beforehand and it also doesn't mean you will be using some Frankenstein product. All of the wine yeasts that are available in those convenient packets were once wild yeasts that got noticed, isolated, and cultured. Learn how to utilize these known microbes to make the cider you want.

AN ARGUMENT FOR CHOOSING YOUR YEAST

In the previous chapter we made a case for going with wild yeasts, captured in a number of creative ways. Now we want to look at the other side and make the argument for choosing your microbe team. We use both approaches, and the dozens of cider batches we do every year on the farm are a mix of wild and cultivated microbial teams. We have also added some cultured wild cards to the deck: kombucha and kefir ciders. We know it's a bit crazy, but it works. It's in this chapter because the kombucha SCOBY (symbiotic culture of bacteria and yeast) is a cellulose mat of both bacteria and yeasts. It's a ready-made microbial colony, and at least some of the bacteria and yeasts are known to have favorable qualities for cider.

One last thing about the life of yeasts — they don't have it easy. Yes, all they have to do is eat and reproduce, but from the very beginning, as their rehydrated selves are poured into your carboy, they have competition from the wild yeasts and bacteria. If you took care of that by sulfiting those wild boys to death, great, but the sulfites linger and are something to be tolerated by the yeasts, and that's not all. Too much sugar can actually be a problem in the beginning for some yeasts, as can an environment that's too hot or too cold. Apples that have high acid levels can cause low pH problems, and then there is the issue of too much alcohol — a rare but possible cause of death for all yeasts. Let's look at the characteristics that both we and the yeasts care about.

Alcohol

Every yeast variety produces alcohol and has a unique tolerance for it. Past their tolerance level, they die. The good news is that this should not be a serious criterion, as nearly all of the commercial yeasts range in a narrow band between 14 and 18 percent ABV, which is some very stiff cider and would only be reached by exceeding the natural sugars in fresh apple juice by either concentrating them or adding additional sugar.

Feeding

Basically, yeasts consume sugar and produce both alcohol and carbon dioxide, but they also need other nutrients, the most important of which is nitrogen. Different strains of yeasts require different amounts of nitrogen. Fresh apple juice has nitrogen, so the question becomes: Does it have enough available nitrogen for the yeasts you have chosen? How can you find out? If you have access to a high-performance liquid chromatography machine, you could learn this and identify every other molecule in your juice, but you probably don't, and neither do we. When we discussed this with commercial cidermakers, we didn't find anyone who regularly tested their juice, but we did find those who knew the accessible nitrogen of the juice that they were buying in bulk.

The other thing we learned is that after a stuck fermentation or two, some of the commercial cidermakers we talked to started using yeast nutrients, like Fermaid A, as an insurance policy of sorts. Our advice is: don't worry about it until you have something go wrong and low nitrogen seems the most viable explanation.

Flavor

In cider, big flavor equates, among other things, to big esters, which are a by-product of fermentation and give us the fruity and floral notes. One characteristic that has been identified in cultured yeast is its ability to enhance the natural esters in apples in the final cider. The two top *S. cerevisiae* strains are *cerevisiae* (yes, it's officially *Saccharomyces cerevisiae cerevisiae*) and *S. cerevisiae bayanus*. If you'd like to enhance the esters in your cider, choose *cerevisiae* over *bayanus*; *bayanus* is neutral on esters, whereas *cerevisiae* does a good job of bringing them out in the cider.

The majority of the acid in apples comes from malic acid, which has a sharp green taste. A few yeasts actually metabolize a small percentage of the malic acid, reducing the overall acidity of the cider. Lactic acid bacteria, like *Oenococcus oeni* and *Lactobaccillus*, can metabolize the majority of malic acid. They wait in the cider until the pH lowers and the ethanol levels rise to a point where they can convert the sharper malic acid to lactic acid, which is more pleasing to the palate. It's not perfect, and sometimes the malolactic fermentation results in less spice and cloves and more horse stall or Band-Aid. Yes, Band-Aid is an official descriptor for tasting ciders, and believe us, it has a distinct smell that you will recognize when you come across it.

Just like the yeasts, bacteria need nutrients, which come from the dead yeasts through a process called autolysis. Basically, once the yeasts die, they release enzymes that consume them. Crazy, right? The result of this autolysis is that the nutrients the yeasts contained are released back into the cider, which is food for the bacteria. It's the cycle of life. This gets back to the importance of the yeasts you select, as some do a better job of autolyzing than others, and they are better food for the lactic acid bacteria.

Temperature

Most yeasts can tolerate a pretty wide range of temperatures — plus or minus 25°F/4°C — and a few have much larger or smaller ranges. Our fermentation caves are built into our farm's hillside and stay between 60°F/16°C and 62°F/17°C year-round, which is on the low end but tolerable for most commercial yeasts. Consider where you will be fermenting and the temperature range of that place. You want to find a cultured yeast that is happy in that range; otherwise it will be stressed, which can lead to it starting slowly (lag phase) and taking a while to build up enough mass to outcompete all the other microbes and dominate the ferment.

Simple Pitched Cider

Simple and pretty bulletproof, this is a great first cider to make. Champagne yeasts are strong fermenters, forgiving of temperature fluctuations, and capable of fermenting all the sugar to a dry finish. When we are teaching beginning pitched cidermaking, we use this recipe.

**YIELD:
3 GALLONS**

½ cup (118 mL) unchlorinated water

1 teaspoon champagne-style yeast (Lalvin EC-1118, Red Star Premier Curvée or Premier Blanc)

3 gallons (11.4 L) preservative-free apple juice

1 (12-ounce/350 mL) bottle dry cider

1. Sanitize a 3-gallon carboy and an airlock with a no-rinse sanitizer.

2. Heat the water to 104°F/40°C and pour into a quart canning jar. Sprinkle the yeast over the hot water, stir gently, and let it sit for 20 minutes. Stir again, measure the temperature of the yeast mixture, and write that down in your cider log or on a piece of masking tape attached to the 3-gallon carboy.

3. Measure the SG of the apple juice with a hydrometer and the temperature with a thermometer and record them in your cider log or on the piece of masking tape attached to the carboy.

4. If the temperature of the yeast and the juice are within 18°F/10°C of each other, proceed to step 5. If not, add ½ cup of the juice to the yeast mixture, gently stir, and wait for 5 minutes. Measure the temperature and if it's within the 18°F/10°C range, move to the next step. If not, add 1 cup more of the apple juice to the yeast mixture, stir, and wait for 5 minutes.

5. Pour the yeast mixture into the carboy. Add some juice to the yeast jar, swirl it around to get all of the yeast remaining in the jar, and pour it into the carboy. Add enough juice to the carboy to fill within 3 or 4 inches of the top. Place a piece of plastic wrap loosely over the opening.

6. Place the carboy on a surface that is easy to clean or set it on a tray or pan. Let it sit in an environment where the temperature is between 55°F/13°C and 65°F/18°C. Bubbles will slowly form after a few days, then build. If the liquid froths out the top and down the sides, simply clean the sides and tray/floor with a wet sponge.

7. When the fireworks are over and the bubbles stay below the top of the carboy (this can take 3 to 10 days or sometimes more), apply the bung or lid, insert the airlock, and fill to the appropriate level with either fresh water or a neutral distilled spirit.

8. The primary fermentation is finished when half to three-quarters of the sugars have been consumed, which you can determine by noting no bubbles being produced or by taking an SG reading. This stage should take between 2 and 3 weeks. If using SG, note your initial SG reading to determine if it's reached the desired level. For example, if the initial SG was 1.060, now it would be 1.030 to 1.015 or lower. Taste the cider used to measure the SG and write any notes in your cider log or on a piece of tape attached to the carboy.

9. Sanitize another 3-gallon carboy, a racking cane, and a siphoning hose.

10. Using a siphon, rack your cider off into the carboy, making sure to draw off all the cider above the lees, without drawing the lees out.

11. Add enough bottled cider to top off the racked cider to within a couple of inches from the top of the new carboy to minimize air contact. Insert the airlock and fill it to the appropriate level with either fresh water or a neutral distilled spirit.

12. Ferment in the same cool environment for at least 1 month or up to 3 months.

13. Take one last SG measurement and calculate your final ABV. Taste, and if you would like it sweeter, back-sweeten it following the technique on page 106.

14. Siphon the cider into clean bottles, secure the tops, and store for at least 1 month before cracking one open. The bottles can be stored in a cool environment out of direct sunlight for a year or more, but at that point the cider will lose some of its sparkle.

From the Cidermaker's Notebook

SINGLE VARIETAL CIDERS FROM COMMON APPLES

Cider apples are popular, as is blending different varieties to make a more complexly flavored cider. However, there is no reason not to make cider if you only have one apple tree, or your local orchard only has culinary apples, or you can get a case of apples on sale at the market. Plus, there is something fun about seeing what an individual variety has to offer so that you know what it will contribute to a collage of flavors.

Here are some popular choices (read: easy to find for the average consumer) for making single-varietal cider:

- Pink Lady
- Cox's Orange Pippin
- Golden Delicious
- Golden Russet
- Jazz

FRUIT CIDERS

Fruit ciders are a natural exploration of flavored ciders. You will notice many of our recipes use the whole fruit in the ferment. This is partly because we like what the wild yeasts from other fruits bring to the table, but it's also because we like working with fresh whole food whenever possible and we grow so much fruit that we are always looking for a good way to preserve it. A downside of using whole fruit in the primary ferment is that some fruits have volatile flavor compounds that are lost in the active fermentation. These flavors literally desert the cider, as they are contained in the CO_2 bubbles that rise up and away, leaving the cider dull and flavorless. Fruit ciders can also be made year-round with frozen puréed fruit or fruit juices that are added to fermented ciders at racking, or even added to fully finished ciders a week before or at bottling.

Cranberry Cider

We have made variations of this recipe using fresh organic cranberries and store-bought apple juice, relying on the natural yeasts in the fresh cranberries to ferment the cider. When it works, it's delicious, but relying on those natural yeasts can be risky. We also make a version that uses a stronger commercial yeast than we typically use to make sure we get a strong fermentation. That seems to work every time and it's simpler, so that's the recipe we have here. Besides being an easy sour cider, it's also a great way to enjoy the two highest antioxidant–rich fruits.

**YIELD:
3 GALLONS**

¼ cup (59 mL) unchlorinated water

1 teaspoon Lalvin 71B or ICV D47 yeast

9 quarts (8.5 L) preservative-free apple juice

3 quarts (2.8 L) preservative-free pure cranberry juice

1 (12-ounce/350 mL) bottle dry cider

1. Sanitize a 3-gallon carboy and an airlock with a no-rinse sanitizer.

2. Heat the water to 104°F/40°C and pour into quart canning jar. Sprinkle the yeast over the hot water, stir gently, and let it sit for 20 minutes. Stir again, measure the temperature of the yeast mixture, and write that down in your cider log or on a piece of masking tape attached to the 3-gallon carboy.

3. Measure the SG of the apple juice with a hydrometer and its temperature with a thermometer and record them in your cider log or on a piece of masking tape attached to the carboy. If the temperature of the yeast and the apple juice are within 18°F/10°C of each other, proceed to step 4. If not, add ½ cup of the apple juice to the yeast mixture, gently stir, and wait for 5 minutes. Measure the temperature and if it's within the 18°F/10°C range, move to the next step. If not, add 1 cup more of the juice, stir, and wait for 5 minutes.

4. Pour the yeast mixture into the carboy. Add some apple juice to the yeast jar, swirl it around to get all of the remaining yeast in the jar, and pour it into the carboy. Add the cranberry juice and enough apple juice to the carboy to fill within 3 to 4 inches of the top. Cover the opening with plastic wrap.

5. Place the carboy on a surface that is easy to clean or set it on a tray or pan. Let it sit in an environment where the temperature is between 55°F/13°C and 65°F/18°C. Bubbles will slowly form after a few days, then build. If the liquid froths out the top and down the sides, simply clean the sides and tray/floor with a wet sponge.

continued on page 185

Cranberry Cider

continued

6. When the bubbles stay below the top of the carboy (this can take 3 to 10 days or sometimes more), apply the bung or lid, insert the airlock, and fill to the appropriate level with either fresh water or a neutral distilled spirit.

7. The primary fermentation is finished when half to three-quarters of the sugars have been consumed, which you can determine by noting no bubbles being produced or by taking an SG reading. This stage should take between 2 and 3 weeks. If using SG, note your initial SG reading to determine if it's reached the desired level. For example, if the initial SG was 1.060, now it would be 1.030 to 1.015 or lower. Taste the cider used to measure the SG and write any tasting notes in your cider log or on a piece of tape attached to the carboy.

8. Sanitize another 3-gallon carboy, a racking cane, and a siphoning hose.

9. Using a siphon, rack your cider off into the carboy, making sure to draw off all the cider above the lees, without drawing the lees out.

10. Add enough bottled cider to top off the racked cider to within a couple of inches from the top of the new carboy to minimize air contact. Insert the airlock and fill it to the appropriate level with either fresh water or a neutral distilled spirit.

11. Ferment in the same cool environment for 1 month.

12. Take one last SG measurement and calculate your final ABV. Taste, and if you would like it sweeter, back-sweeten it following the technique on page 106.

13. Siphon the cider into clean bottles, secure the tops, and store for at least 1 month before cracking one open. The bottles can be stored in a cool environment out of direct sunlight for a year or more, but at that point the cider will lose some of its sparkle.

Blair Smith of Apple Outlaw Cidery

Everyone has their own path to starting a business. A high percentage of the small-scale commercial artisan cidermakers we have talked with started their journey with a passion for the libation — often a hobby gone big, which requires sourcing apples or juice on a larger scale. But the Smith family just wanted to move to the country. They were looking to trade the high-tech Silicon Valley life for one in which they could be surrounded by more trees than people. They were searching in the Applegate Valley of southern Oregon (yes, they are our neighbors), and at that time, the perfect home happened to come with a mature apple orchard. That wasn't the attraction, nor did they know enough about what they were getting themselves into to have it be a detraction either.

A year or so after moving there, they wanted to maintain the orchard, which at first meant bringing it back from the near neglect state it was in when they'd moved there. They pruned; fixed fences to keep the bears, racoons, and other "outlaws" out; they pulled out blackberry vines that were threatening to take over. Blair (head cidermaker and orchardist) said that simply maintaining the orchard cost money, so they decided to take the fruit to the market to cover costs. Fresh apples weren't profitable, especially since at that point the trees were a mix of no-longer-fashionable apples — a majority

of them being Red Delicious — so they contracted with a local juice maker to press their apples into fresh sweet cider. The juice company was nearing the end of its run, and by the next year it asked the Smiths if they would like to purchase the juice pressing equipment. The fresh apple juice had, as hoped, added value to the apples that came off the orchard, so the Smiths bought the equipment and became an organic juice company.

Our boys were young teenagers when the Smiths hired them to work at their pressing operation. The boys would ride their bikes up the mountain in the morning, then come back well past dark on those late fall days. Kirsten admits she remembers it well, as she began to wonder, okay worry, slightly on those late nights. They were always fine. The

Smiths and the boys worked all day, fueled by lots of peanut butter-and-jelly sandwiches. By the time our youngest son left the pressing job, the work was smooth and they were pressing many times the number of apples they had in their first season. It's all about finding your groove.

Still, the value added didn't add enough to make the orchard pay for itself, let alone be profitable. So Blair focused on quality and flavors people wanted. They bought bigger tanks and lots of oak barrels, and they ramped up their fledgling cider business. They soon landed coveted awards as a new cidery and the attention of buyers as far away as Japan. Apple Outlaw is now a thriving family-owned and -operated cidery that is a fixture in the Applegate Valley of Oregon.

Stone Fruit Cider

In our forest garden we have plum, cherry, and peach trees that produce every year — to varying degrees — and one apricot tree that is usually either dying, dead, or newly replaced and has yet to know its fate. We used to dry most of this fruit to get the kids through snack time in gray, rainy winters, but now that they are grown, we just pit the fruits and throw them in the freezer until apple-pressing time. For a special batch we add whatever we find in the freezer, and come springtime we have something that reminds us of last autumn on the farm. While it is tempting to chuck the fruit in the freezer whole, depending upon the fruit, it can make pitting a little harder. (See the box on why to pit stone fruit on page 62.)

In this recipe we prefer the K1-V1116 yeast (*S. cerevisiae*) because it does well at lower temperatures and is a strong floral ester producer that is well matched for the additional stone fruit.

YIELD: 3 GALLONS

- ¼ cup (59 mL) unchlorinated water
- 1 gallon (3.8 L) fresh or frozen pitted stone fruit
- 2 gallons (7.6 L) preservative-free apple juice
- 1 teaspoon Lalvin K1-V1116 yeast
- 1½ teaspoons (7.5 g) pectic enzyme powder
- 1 (12-ounce/350 mL) bottle dry cider

1. Sanitize a 3-gallon carboy and an airlock with a no-rinse sanitizer.

2. Heat the water to 104°F/40°C and pour into a quart canning jar. Sprinkle the yeast over the hot water, stir gently, and let it sit for 20 minutes. Stir again, measure the temperature of the yeast mixture, and write that down in your cider log or on a piece of masking tape attached to the 3-gallon carboy.

3. Mash the fruit by hand or run it through a food processor or blender to get something between a coarse chop and a purée. You might be wondering why we don't run these fruits through the grinder and press with the apples, but with our press the cloths get gunked up quickly. It's easier to process them separately.

4. In a blender or food processor, purée ½ cup of the mashed fruit with 1 cup of apple juice. Measure the SG of this mix with a hydrometer and record it in your cider log or on a piece of masking tape attached to the 3-gallon carboy.

5. Measure the SG of the apple juice with a hydrometer and the temperature with the thermometer and record them in your cider log or on a piece of masking tape attached to the carboy. If the temperature of the yeast and the juice are within 18°F/10°C of each other, proceed to step 6. If not, add ½ cup of the apple juice to the yeast mixture, gently stir, and wait for 5 minutes.

Measure the temperature and if it's within the 18°F/10°C range, move to the next step. If not, add 1 cup more of the juice, stir, and wait for 5 minutes.

6. Pour the yeast mixture into the carboy. Add some apple juice to the yeast jar, swirl it around to get all of the remaining yeast in the jar, and pour it into the carboy. Add the stone fruit mash and enough apple juice to the carboy to fill within 3 to 4 inches of the top. Cover the top with plastic wrap.

7. Place the carboy on a surface that is easy to clean or set it on a tray or pan. Let it sit in an environment where the temperature is between 55°F/13°C and 65°F/18°C. Bubbles will slowly form after a few days, then build. If the liquid froths out the top and down the sides, simply clean the sides and tray/floor with a wet sponge.

8. When the bubbles stay below the top of carboy, apply the bung or lid, insert the airlock, and fill to the appropriate level with either fresh water or a neutral distilled spirit. Depending upon temperature, this can take 3 to 10 days or sometimes more.

9. The primary fermentation is finished when half to three-quarters of the sugars have been consumed, which you can determine by noting no bubbles being produced or by taking an SG reading. This stage should take between 2 and 3 weeks. If using SG, note your initial SG reading to determine if it's reached the desired level. For example, if the initial SG was 1.060, now it would be 1.030 to 1.015 or lower. Taste the cider used to measure the SG and write any tasting notes in your cider log or on a piece of tape attached to the carboy.

10. Sanitize another 3-gallon carboy, a racking cane, and a siphoning hose.

11. Using a siphon, rack your cider off into the carboy, making sure to draw off all the cider above the lees, without drawing the lees out.

12. Add enough bottled cider to top off the racked cider to within a couple of inches from the top of the new carboy to minimize air contact. Insert the airlock and fill it to the appropriate level with either fresh water or a neutral distilled spirit.

13. Ferment in the same cool environment for 1 month or more.

14. Take one last SG measurement and calculate your final ABV. Taste, and if you would like it sweeter, back-sweeten it following the technique on page 106.

15. Siphon the cider into clean bottles, secure the tops, and store for at least 1 month before cracking one open. The bottles can be stored in a cool environment out of direct sunlight for a year or more, but at that point the cider will lose some of its sparkle.

Lyndon Smith of Botanist and Barrel

We met Lyndon Smith, cofounder of Botanist and Barrel Cidery and Winery and maker of natural Southern ciders, at the Western North Carolina Fermenting Festival. Since 2017, Lyndon, his wife, sister, and brother-in-law have created natural ciders and wines using old-world methods with a modern spin. Their practices are rooted in a mix of passion for fermentation and farming the land; their blueberry farm provides the anchor and inspiration. Their ciders and wines are made by capturing flavors from the ingredients in a 200 mile radius of Cedar Grove, North Carolina. They are wild, unfiltered, unfined, unpasteurized, and fermented dry. Soon into our conversation Lyndon said sweet ciders "aren't what nature intends." He explained, "There is nothing about the process that will leave any sugars behind." Lyndon manages the fruits and sugars to achieve both still and sparkling (using *pét-nat*) ciders that are full bodied and delicious.

Since the Botanist and Barrel team also makes grape and fruit wines, they often use their leftover skins — like Montmorency cherries, which were used to make the bottle of Skin Contact cider that blew our minds. Lyndon generously shared how to layer on flavor with some wine techniques, which we felt compelled to include for those of you looking to dabble in some other techniques. The basic idea is that skins (of already pressed fruit) are added to wild fermenting cider just at the end of primary fermentation. He said that adding skins then, when the yeast is hungry, rejuvenates the fermentation and creates a deeper, richer, integrated, and balanced cider/wine.

Lyndon explained that as they looked at the history of winemaking, they saw how piquettes (low-alcohol wine made from pomace — the ciderkin of wine), rosés, and reds are made with varying levels of skin contact. "We wanted to make a blueberry wine, in the likeness of a young Beaujolais, so we pressed off the juice and rested it on its skins for a few days. The skins had so much life left in them that we made a blueberry piquette." The team at Botanist and Barrel transfers ideas and sometimes materials from their wine experiments to cidermaking. For example, when they were making a bone-dry cherry wine and noticed the juice itself had plenty of tannin and color, they didn't allow any additional skin contact in order to keep it balanced. Instead, they got the idea to add the skins to cider. They loved it so much they started doing this with a variety of fruits like pineapple; pinot gris, muscadine, traminette, and mourvedre grapes; and, of course, apples.

New England–Style Cider

This style of cider is made with apples of the Northeast, like Northern Spy, Roxbury Russet, and Baldwin, which give the cider a higher level of acidity, flavor nuances, and increased alcohol (which can be substantial). Sugars, molasses, honey, and raisins can all be employed to boost the sugar-to-alcohol process, resulting in a cider that clocks in between 7 and 13 percent alcohol.

For a wilder version that tries to get as close as possible to what a traditional farmhouse version would taste like, use the same ingredients but follow the process for making a wild cider yeast in chapter 6. This recipe has less funk and sour than a wild version, and instead brings out other unique traits of this style: the brown sugar and molasses notes and a little nutrient and tannic kick from the raisins.

YIELD: 3 GALLONS

---

3 gallons (11.4 L) pasteurized apple juice

1¼ cups (296 mL) unchlorinated water

¾ cup (160 g) organic brown sugar

1 teaspoon Lalvin ICV 71B yeast

1 cup (130 g) organic raisins

¾ cup (255 g) unsulfured molasses

1 cup (41 g) oak chips

---

1. Sanitize a 3-gallon carboy and an airlock with a no-rinse sanitizer.

2. Measure the SG of the juice with a hydrometer and the temperature with a thermometer and record them in your cider log or on a piece of masking tape attached to the carboy.

3. In a saucepan over medium-high heat, mix together ¾ cup of the water and the sugar, stirring continuously until the sugar dissolves completely to create a simple syrup. Set aside to cool to room temperature.

4. Pour the cooled simple syrup into the carboy.

5. Heat the remaining ½ cup water to 104°F/40°C and pour into a quart canning jar. Sprinkle the yeast over the hot water, stir gently, and let it sit for 20 minutes. Stir again, measure the temperature of the yeast mixture, and write that down in your cider log or on a piece of masking tape attached to the 3-gallon carboy.

6. If the temperature of the yeast and the juice are within 18°F/10°C of each other, proceed to step 7. If not, add ½ cup of the juice to the yeast mixture, gently stir, and wait for 5 minutes. Measure the temperature and if it's within the 18°F/10°C range, move to the next step. If not add 1 cup more of the juice, stir, and wait for 5 minutes.

continued on next page

7. Pour the yeast mixture into the carboy with the simple syrup. Add some juice to the yeast jar, swirl it around to get all the yeast remaining in the jar, and pour it in the carboy. Add enough juice to the carboy to fill within 3 to 4 inches of the top. Cover the opening with plastic wrap. Reserve at least 1 quart of the apple juice, secure with a lid, and refrigerate for later use.

8. Place the carboy on a surface that is easy to clean or set it on a tray or pan. Let it sit in an environment where the temperature is between 55°F/13°C and 65°F/18°C. Bubbles will slowly form after a few days, then build. If the liquid froths out the top and down the sides, simply clean the sides and tray/floor with a wet sponge.

9. When the bubbles stay below the top of carboy, apply the bung or lid, insert the airlock, and fill with either fresh water or a neutral distilled spirit. Depending upon the yeasts, the temperature, and the juice, this can take 3 to 10 days or sometimes more.

10. The primary fermentation is finished when half to three-quarters of the sugars have been consumed, which you can determine by noting no bubbles being produced or by taking an SG reading. This stage should take between 2 and 3 weeks. If using SG, note your initial SG reading to determine if it's reached the desired level, which should be around 1.020. Taste the cider used to measure the SG and write any notes in your cider log or on a piece of masking tape attached to the carboy.

11. In a saucepan over medium-high heat, mix together the raisins, molasses, and ½ cup of the reserved apple juice. When the mixture reaches the boiling point, remove the saucepan from the heat and stir for 1 minute. Let cool to room temperature.

12. Remove the airlock from the carboy and add the molasses-raisin mixture to the cider. Replace the airlock.

13. Ferment in the same cool environment until it reaches an SG of 1.000.

14. Sanitize another 3-gallon carboy, a racking cane, and a siphoning hose.

15. Using a siphon, rack your cider off into the carboy.

16. In a saucepan over low heat, mix together the remaining apple juice and the oak chips and simmer for 15 minutes, stirring occasionally to incorporate the chips. Let cool and taste. You should pick up a pronounced tannic astringency. Strain, and add the juice to the new carboy to fill to within a couple of inches from the top to minimize air contact. Insert the airlock and fill to the appropriate level with either fresh water or a neutral distilled spirit. If you have remaining oaked juice, you can freeze it for future cider batches. If you didn't taste as much oak as you had hoped, save the chips by spreading them on a clean cloth to dry and storing in an airtight carboy at room temperature. You will have a second chance to add oak flavor near the end of fermentation.

17. Ferment in the same cool environment for at least 1 month or up to 3 months. About 1 week to 10 days before the end of fermentation, draw a sample and taste. If you don't pick up as much oak as you had hoped, add the reserved oak chips (see Barrel Aging without the Barrel, page 103) to the container and soak for 7 to 10 days.

18. Take one last SG measurement and calculate your final ABV. Taste, and if you would like it sweeter, back-sweeten it following the technique on page 106.

19. Siphon the cider into clean bottles, secure the tops, and store for at least 4 months before cracking one open. The bottles can be stored in a cool environment out of direct sunlight for a year or more.

Black Currant

Fresh black currants aren't always readily available, but for those of you that have access to some, this makes a beautiful, deep maroon cider. We used a ratio of 1 quart of mashed black currant to 1 gallon of pasteurized apple juice and pitched it with Lalvin ICV D47 yeast. An interesting note for this cider is that early in the making, it had nice floral undertones, but its flavor and mouthfeel were thin. We decided to keep the cider on the lees, stirring it regularly. Just a few weeks in, the cider developed a full, round mouthfeel and the fruity floral flavors started to come forward. Bolstered by this improvement we continued, and honestly it was never as good as it once was — so watch yours closely and when it amazes, it's time to bottle it.

Gooseberry Cider

This is one of our favorite ciders. If you grow gooseberries or, like us, have a friend who grows them, you should try a simple gooseberry cider. We try to make it every year and we haven't yet been disappointed. We use a pound of berries for every gallon of juice. If you use

fresh juice — even if your berries were frozen — it will make a nice wild cider with lots of fruit and a predominate sour currant note. If you are using purchased juice and frozen berries, pitch with Lalvin EC-1118 or another champagne-type yeast.

Blackberry Cider

In our neck of the woods, the Himalayan blackberry is a weed. Blackberries, like apples, are a big part of our late-summer menu — we enjoy them in pies, cobblers, and jam. They also have a regular spot in the carboys in the cave, making blackberry wine but more often cider that has fermented with extra sugar added on day one, or chaptalized through the primary fermentation. Christopher's favorite ratio is 2 quarts of mashed blackberries to 3 gallons of fresh unpasteurized juice, with 3 pounds of dark, demerara-type sugar. This is usually a wild ferment, but when using commercial yeast, we use Red Star Côtes des Blancs.

LIGHTLY
FERMENTED
SOFTISH
CIDERS

The following are lighthearted sweet apple-juicy ciders. The kombucha and kefir ciders are kid-friendly with a low alcohol content (well under 1 percent) and can be another way to reduce sugar intake and add probiotics to your diet. We've made some of these fermented apple beverages into summer apple ice pops. The light fermentation is perfect for a little effervescence and a little funk. We make these for drinking within a week or two; they are not meant for long-term bottling.

Ciderkin, Fermented or Not

Also called "water cider," this is the ultimate zero-waste product. It is made by steeping leftover pomace in boiling water, producing an "apple tea" or ciderkin that is a lighter apple juice that can be fermented. This low-alcohol cider was traditionally made to be drunk by the children, back when the water wasn't safe. The fermentation took care of the unsafe microbes, and did not result in drunk babies. The ciderkin was fermented naturally with time — to do this, simply keep it in the fridge and taste it as it gets funky. You can also add molasses, as they did back in the day, to add a rich flavor and boost the alcohol content.

Because the low alcohol content, when fermented, doesn't provide long-term preservation, this is a fun cider to drink soon after pressing to help bide the time before your regular cider is ready. Molasses, light or dark, is 60 percent sugar, so adding a touch over 1 cup to a quart of water will produce a cider with 6 percent ABV. You can use these numbers to play with this recipe. You will need recycled plastic bottles for this cider.

**YIELD:
2-3 GALLONS**

Pomace to fill a sanitized 5-gallon food-grade bucket ½–¾ full

About 3 gallons (11.4 L) boiling water

1 cup (340 g) unsulphured molasses (optional)

½ cup (118 mL) water/ciderkin juice

1 teaspoon champagne yeast (optional)

1. Sanitize a 5-gallon bucket with a no-rinse sanitizer.

2. Put the pomace into the bucket and add enough boiling water to fill. Cover and leave overnight.

3. The next morning, place this wet pomace in the press. The juice that comes out is the ciderkin. It can be enjoyed as is for a lighter sweet cider. If that is your goal, place in jugs or jars and refrigerate. It will last 7 to 10 days.

4. If fermenting the ciderkin, put it in a bucket. Skip to step 6 if not using molasses for a wild, lower-alcohol ferment.

5. If using molasses, warm 2 cups of the ciderkin in a medium saucepan over medium heat until hot but not boiling, about 160°F/71°C. Combine the molasses and warm ciderkin in a medium bowl, stirring until the molasses is dissolved. Pour the mixture into the bucket with the rest of the ciderkin and stir to incorporate.

6. Heat the water or ciderkin juice to 104°F/40°C and pour into a pint canning jar. Sprinkle the yeast over the hot water, stir gently, and let it sit for 20 minutes. Meanwhile, wash and rinse your recycled plastic bottles.

7.   Fill one bottle about one-third full with no-rinse sanitizer, close the cap, and shake for 20 seconds. You can pour the sanitizer into your next bottle and reuse three or four times. Invert the bottles on a dish rack to dry.

8.   Stir the cider again. Fill the dry bottles with cider to about 2 inches from the tops and screw on the caps. Since the carbon dioxide has nowhere to escape, the bottle will begin to expand. For that reason, we suggest you leave this bottle on your counter in plain sight, in a cool corner away from the heat of major appliances. Remember that wild yeasts like it on the cooler side.

9.   Depending upon the size of your bottles, the activity of yeasts in your fresh juice, and the temperature, you could see the bottle swelling in a few days to a week or more. You are going to need to "burp" this cider by partially unscrewing the cap, letting the gas escape, then quickly screwing it back tight. You may end up doing this daily or twice a day, depending upon the amount of yeast activity and carbon dioxide being created. Let it sit for a week if you want it on the sweeter side, or longer if you want it drier.

10.   You can refrigerate this sparkling cider but it will continue to ferment even in the refrigerator, so check in on it every few days and give it a wee burp.

Kombucha Cider

This is like a tangy, fizzy apple juice with a lot less sugar and a healthy shot of probiotics and strong digestive enzymes. While the literature says the alcohol content can come in between 0.5 and 1 percent, ours has never come in over a very low 0.13.

Christopher was the first to think it would be fun and tasty to put a kombucha SCOBY on apple juice, but it took Kirsten a while to come around to the idea. As a vinegar maker, she (wrongly) assumed it would merely turn the juice into vinegar, but after we did further research, we decided to try. The resulting kombucha cider was delicious. It really is its own thing — sweet and sour, a little effervescent, and still very appley. We played around with using other fruit juices, but plain apple juice or sweet cider is our favorite. We prefer the tang after 5 to 9 days of fermentation (depending on conditions); much longer and it starts tasting decidedly like vinegar. You will likely have your own idea of a sweet spot.

YIELD: 1 GALLON

1 gallon minus 2 cups (3.5 liters) good-quality unfiltered apple juice, or fresh-pressed sweet cider

1½ cups (355 mL) mature kombucha for starter

1 (4–5 ounces/115–140 g) SCOBY, or a piece that will cover at least 25 percent of the surface

1. Pour the apple juice into a sanitized 1-gallon jar or brewing vessel with an open top. Add the kombucha starter to acidify the cider and give it a microbial jolt. This is called backslopping.

2. Using clean hands (no antibacterial soap) or clean rubber tongs, or wearing rubber gloves, carefully place the SCOBY on top of the cider. It is okay if it sinks.

3. Cover the jar with a clean piece of muslin, woven cloth, or a round coffee filter and secure with a rubber band.

4. Place the jar in a spot that is slightly warmer than room temperature (78°F/26°C to 80°F/27°C). Ferment for 5 to 10 days.

5. Begin tasting on day 5. If the flavor is pleasing, remove the SCOBY. If it doesn't have the right balance of acidity and sweetness, continue to ferment, tasting it every day. (Note: This is now a cider SCOBY, as it has begun to adjust its microbial community to process apple juice. If you want to save your cider SCOBY for another batch, be sure to store it in an airtight jar, fully submerged in some of the fermented liquid. The liquid will also serve as your starter when you make the next batch.)

6. Store the kombucha cider in the refrigerator and drink within 1 week. If you want a lot of carbonation, pour the kombucha cider into small sanitized plastic soda bottles. These will fill quickly with bubbles. Open slowly over the sink.

KOMBUCHA SCOBYS

Kombucha hardly needs an introduction in today's world. It's that lightly carbonated fermented tea traditionally made with sweetened black tea. Kombucha is produced by a cellulose blob called a pellicle, zoogleal mat, or most often a SCOBY (symbiotic culture of bacteria and yeast), which, just as it sounds, is a community of microbes helping each other out. Generally speaking, the yeasts (mainly *Zygosaccharomyces* with *Saccharomyces cerevisiae* and *Brettanomyces* close behind) consume the sugar and turn it into alcohol, with the acetic bacteria (*Gluconacetobacter* and *Acetobacter* in order of abundance) waiting in the wings to consume the alcohol and turn it into acid. It is, as with most everything in life, much more complicated than this — the microbial members of the SCOBY differ by geographical location as well as the type of tea and sugar used. For example, a study found that some SCOBYs produced kombucha that was high in lactic acid bacteria, whereas other kombuchas have none.

Because a kombucha SCOBY always needs to be fed, you will use bottled apple juice most of the year. Once your SCOBY is a community of microbes fit for processing apple juice, you will want to keep feeding it apple juice. When not in use, refrigerate the SCOBY, making sure it is fully submerged in its liquid.

Kefir-Cultured Cider

We loved the flavor of this ferment. The drink is rich, with a creamy mouthfeel and a light acidity, but it is in no way sour like kombucha. There is also less of a "sugary" taste behind the tang, as there is in the kombucha cider. It has good, sparkly effervescence and can taste a little funky in that good fermenty way. Kefir fermentation can produce a small amount of alcohol (0.5 to 1 percent).

We make this in either a 1½-quart recycled plastic bottle or a 1-quart glass milk bottle. We like the narrow neck because we can easily cap it for a few hours before drinking in order to create effervescence. Additionally, we have found that if you add new juice when you have 1 or 2 cups left in the bottle, it will ferment again. One packet of culture can make multiple batches.

YIELD: 1-2 QUARTS

1-2 quarts (1-2 L) pasteurized apple juice

1 teaspoon powdered kefir starter culture

1. Sanitize a plastic or glass bottle with a no-rinse sanitizer.

2. Add enough juice to the bottle to fill within about 4 inches from the top. Add the powdered kefir starter, secure with the lid, and shake to combine. Open the lid and add enough juice to fill within about 2 inches from the top. Secure the lid.

3. Since the carbon dioxide has nowhere to escape, the bottle will start to expand. For that reason, we suggest you leave this bottle on your counter in plain sight, in a cool corner away from the heat of major appliances. Remember that wild yeasts like it on the cooler side.

4. Depending upon the size of your bottle, the activity of yeasts in your fresh juice, and the temperature, you should see the bottle swelling in a few days to a week or more. You are going to need to quickly "burp" this cider by partially unscrewing the cap, letting the gas escape, then quickly screwing it back on tight. (If you use a glass milk bottle, it won't expand, but the top will pop off if there is too much pressure, so it still needs to be burped.) You may end up doing this daily or twice a day, depending upon the amount of yeast activity and carbon dioxide being created. Let it sit for a week if you want it on the sweeter side, or longer if you want it drier.

5. The idea is to enjoy this cider while it still has some sweetness and a nice amount of gas so it's naturally sparkling. Refrigerate or keep out on the counter at room temperature for a few days. When there is one-quarter of the cider left in the bottle, add more juice and in a day or two, you will have more kefir cider.

KEFIR GRAINS

We were amazed at how different kefir-cultured cider tasted from kombucha cider — so much so that we offered a blind taste test to everyone who came by. In our small sampling, the kefir cider won almost every time.

The kefir microbial community contains bacteria and yeast, like SCOBYs, but forms in "grains" instead of a mat. Kefir grains are more delicate than kombucha SCOBYs. They are vulnerable to changes in temperature and can starve if they run out of sugar.

There are two kinds of grains: milk kefir grains that thrive by consuming lactose (milk sugar) and water kefir grains that use sucrose as their fuel. Milk kefir is believed to have originated in the Caucasus Mountains, while water kefir possibly comes from Mexico, where there is historical evidence of fermented prickly pear cactus in a drink called *tipicos*. We tried both and found them to be successful, though surprisingly, the milk

grains seemed to do a better job. Once you have used them for cider for a few batches, you must "refresh" them by letting them process a batch of milk. Rinse the milk off gently when ready to put them back into the apple juice. The grains work fast — the kefir-fermented cider is ready to drink in as little as 2 days — but the grains need to be maintained if you want to reuse them. This works well if you always plan to have some kefir cider going, but it can be a project to keep up with it.

The "single use" powered cultures are a collection of 7 to 9 strains of bacteria (dependent on brand), whereas the grains will have many different yeast strains, bacteria, and subspecies as well, so there is more microbial bang for sure. In the Kefir-Cultured Cider recipe, we use the powdered packets that don't require upkeep. After all, you are cidermakers, and we want to keep it simple. However, as we mentioned, the regular grains work well: feel free to use them.

CRAFTING CIDER WITH WILD BOTANICAL YEASTS

There are many ways to capture the essence, aromas, and flavors of your local landscape. In this chapter and the next, we explain the ways you can do this with or without local apples. Even if you only have pasteurized cider, you can still embark on a wild cider adventure with local yeasts and flavors. You'll learn how to harvest the wild yeasts from flowers and botanicals, then use those yeasts to ferment pasteurized apple juice. You may get to know your town or city in ways you never imagined as you seek out little patches of nature near you — not just blossoms but leaves, bark, sap, seeds, stems, and roots. It's all about fun, experimentation, and getting to know your local flavorshed in completely new ways; it is not about predictable results. Embrace the variations. In the next chapter, we talk about how to infuse already fermented cider with botanicals.

BLOSSOM AND BOTANICAL WILD YEASTS

Wild yeasts are everywhere, and they are especially prevalent on blossoms. You can have a lot of fun making small batches of wild ciders year-round with seasonal blossoms, botanicals, and other fruit skins. In many cases, you will get a wonderful cider that has the essence of your yeast source. Due to the unpredictable nature of wild yeasts, you can discover untamed flavors — and as you can imagine, some are better than others, but the good ones knock it out of the park. In this section we share with you some of our maker notes, ideas, fails, and some recipes.

As a maker, you have two choices: you can just go for it and drop your wild yeast host (botanicals or blossoms) into the pasteurized juice that you intend to ferment, or you can make a wild yeast starter. We give you instructions for both. Is there an advantage to one or the other? It's easy to drop your host material into the apple juice, and your resulting cider may have a stronger essence of the blossom or botanical, but it's a gamble, as you have less control over any possible contaminating microbes that may hitch a ride on the plant matter. If you don't like the results, you have a chance to abort the experiment before you end up with a large failed batch. A wild yeast starter will give you more control, as you are able to discover any spoiling characters in the pint jar before committing to a whole gallon of juice or more.

We found, for example, that there are seasonal differences in blossoms that had significant impact on the flavor of our ciders. The early spring blossoms — like wild violet, dandelion, early rose, and plum, for instance, added only their yeast and aromas with no extra funk. We use handfuls of these and other early blossoms directly in our fermenting vessels. As the days got warmer and longer, we experienced more bacterial and sour flavors. A whole carboy of infused elderberry blossom cider resulted in a viscous ropey mess (see Troubleshooting, page 301). One batch with linden blossoms made a nice sour with floral notes, but some might feel it had been overly influenced by bacteria — taste is subjective. It is possible the yeasts weren't very strong. Either way, we could have caught this in a couple of cups of yeast starter juice instead of risking a few gallons. We found as the weather cools again in fall, the success rate rises. This is when we use the powdery dusting on fruit and late garden flowers, like sunflower petals and Thai basil flowers.

THE "CRADLE-OF-YEAST" IS LIKELY IN CHINA

Yeast can be found throughout the world — in human-driven creations such as bread, wine, sake, and cider, but it's also in plants and soil and on insects. Two geneticists, Gianni Liti from the Université Côte d'Azur and Joseph Schacherer from the Université de Strasbourg, set out to sequence the genomes of 1,000 yeast strains. They looked under rocks, in tree bark, at human infections, in termite mounds, and anywhere they, or anyone else, could think of from around in the world. It took them about 5 years.

Their sequencing project pointed toward something that scientists had already suspected: yeast probably originated in China. China contained the most genetically diverse array of yeasts than anywhere else. In other words, there were more differences in wild yeasts found in different parts of China than there were between wild North American and European yeasts. This makes sense if you consider that all yeasts (wild or domesticated) that made it out of Asia were likely carried and spread by the movement of flora and fauna and human domestication — those tasty breads and beverages.

Foraging for Yeast

Perhaps you are a wildcrafter at heart, or when given the option you will always choose the most nature-based path, or maybe you just like the hunt for new flavor. If any of these sound familiar, give yeast foraging a try.

Yeasts eat sugar, so it stands to reason they are found most readily on blossoms and the fruiting bodies of the plant. The white powdery bloom that you see on many fruits — like grapes, blueberries, plums, and even prickly pear or juniper berries — is loaded with yeast.

As foraging or wildcrafting has become more popular, the concern is whether it is sustainable. Harvest wild edibles with care. As you discover the plants nearby, you can learn how to propagate the ones you love. Often, you can find groups of people interested in native plants to help you do this. If you have space, you can also plant your favorites in your garden. Giving back to the landscape can be as simple as heading out in fall and collecting seeds for the purpose of replanting in the wild. Don't forget that harvesting nonnative invasive plants can be a good thing. For instance, if you live in the Pacific Northwest, where Himalayan blackberries eat buildings, go ahead and make as much wild blackberry cider as you want.

When you are foraging for yeast, you don't need to harvest a huge quantity (take less than half of any one given plant, and less if there is a small population) and the plants don't need to be "wild" for the yeast to be. Much of what we pick are domesticated plants — for example, the petals of sunflowers in our garden or blossoms from our fruit trees. (And if you want to get on your science nerd, or do fun yeast science with kids, try isolating yeasts in petri dishes; see Suppliers, page 321 for information about how to get started.) Here are a few important things to keep in mind while foraging.

- Avoid plants growing by roadsides, as they will have collected exhaust and may bring harmful chemicals into the mix. By the same token, don't harvest from plants, trees, or blossoms that have been sprayed with pesticides or herbicides. In short, spend some time observing the areas you are interested in and make sure the plants aren't contaminated.

- Do your research. Know what plants are rare and shouldn't be harvested at all. Don't harvest from protected and preserve areas.

- Not every flower (or plant) is edible. Again, do your research and be sure that you identify the plant and flower and only use the edible parts.

- Use flowers and botanicals sparingly — at first, especially — as some may cause digestive upset in some people.

Apple blossoms (*Malus pumila*). These are the incubator for the yeasts that the apples carry in fall, so of course we wanted to see what would happen when we used the blossoms in cider. The yeasts are vigorous, especially from our old fuchsia-colored flowering crabapple. The cider is delicious, though we don't detect any particularly floral flavor like we do with plum blossoms. Honestly, these ciders taste exactly like our fall wild-fermented ciders; this is handy to know if you run out in spring — make a little more from purchased pasteurized juice. If you want to use apple blossom yeast in fall for pasteurized, freshly pressed juice, you could make a wild yeast starter (page 211) in spring and keep it going for use later in the season. For the cider, follow the recipe for Plum Blossom Cider (page 217).

Common juniper (*Juniperus communis*). These berries are best known for their role in gin and are actually modified cones. There are approximately 40 species of juniper, but only a few are edible and of those, only one is used culinarily; some are quite bitter, and a small number are poisonous. If you want to use wild ones for a flavorful yeast starter, check and identify your local varieties before picking. These make an active wild yeast starter, adding wonderful aromas to cider.

Dandelions (*Taraxacum officinale*). Dandelion wine meets cider. The wild yeasts on dandelion blossoms are perfect for starting a spring cider from pasteurized juice. Use around 35 grams of the flower per gallon of juice. If you don't want bitter compounds, use just the petals plucked from the milky green sepal, but note that it is also these bitter compounds that give more depth to the juice as it ferments. We have also made a delicious clear sparkling cider with a combination of dandelion flowers and lemon balm (*Melissa officinalis*). We combined 35 grams of lemon balm with 30 grams of dandelion blossoms in fresh pasteurized juice using champagne yeast. It was citrusy when we bottled it, and after 11 months of conditioning with a carbon drop, it was bright, nearly clear in color, with a clean green apple flavor, and a hint of citrus and light herbs. We also make infusions with the roots (see notes on page 236).

Garden sage blossoms (*Salvia officinalis*). When the purple florals of sage filled the herb garden in spring, one of us couldn't help but try capturing the yeasts on the blossoms by stuffing a handful of blossoms into a gallon of pasteurized juice to make cider. If a cider with floral Thanksgiving stuffing aromas doesn't sound appealing, wait — it made the most amazing vinegar.

Lilac (*Syringa vulgaris*). Lilacs have always had a special significance to Kirsten; there was a lilac bush wafting its perfume outside the window of one of her childhood homes, and it left an impression. We wanted to capture the full breadth of their cheery, purple spring energy so we used them in juice to capture the wild yeast and as much of the flavor as we could. We use anywhere from 2 to 3 cups of blossoms (plucked from the stems) for each 1 gallon of juice. The resulting ciders are delightful and easy drinking. They finish at an SG of 1.000 but still have a bit of residual sweetness. They are floral, as you would imagine, but just lightly (not like your great-grandmother's perfume). Finally, they have a lovely, light tannic bitterness, and we found the white lilacs are slightly tastier than the purple.

We were also lucky enough to have tasted WildCraft's lilac cider, which is made in a completely different way—a blend of cider and lilac wine. Sean Kelly

(page 230) makes this community cider as a fund-raiser. Community members gather the lilacs, which he makes into a lilac wine. The cider is blended with 10 percent of this wine. "Plant sugars can take years to ferment. Flowers are full of sugars that are perceived by smell but difficult to taste because of the many bitter constituents. However, if one were to juice the flowers, there would be more sugars present in the juice than from that of many of the apples that we pressed last year," said Sean.

Avoid the narrow-leaf variety of milkweed (right).

Thai basil (*Ocimum basilicum* var. *thyrsiflora*). Neither one of us is a huge fan of anise or licorice, though we love this Southeast Asian basil, which has those qualities. Thai basil gives cider some nice spice notes but it doesn't taste like you are bringing the savory kitchen into your cider (see garden sage on previous page). The slight licorice qualities of Thai basil carry through in a familiar, recognizable way. People that taste it feel like they know it but can't place it — it's on the tip of their tongue. We use Thai basil as a wild blossom yeast using a starter jar.

Tulsi/holy basil (*Ocimum tenuiflorum*). This herb is in the basil family and we grow it in our garden. Two of our favorite homemade ciders include tulsi. We make a wild yeast starter with the long gangly blossoms, then keep feeding it until we are ready to make the cider. It makes a sour cider. Five flower stalks are enough for at least a gallon or two of cider. The other way we use it is in the tea form. Our friends at Oshala Farm make an herbal tisane called Be Good to Yourself, which is a mixture of tulsi, nettle, and borage flower and leaf, which you will read about on page 238. It is a delicate, calming cider.

Wild violets (*Viola* spp.). You can read all about these harbingers of spring on page 250 as well as find a recipe to use them to infuse cider. These can also be used for their wild yeasts. The flowers make an active ferment.

Milkweed blossoms (*Asclepias syriaca*). Common milkweed has an alluring, sweet, lightly spicy aroma all its own that reminds Kirsten of honey. No wonder the pollinators adore it! This plant is the only thing that monarch caterpillars eat, so we grow a lot of it on our hillside homestead. We don't harvest many flowers for our own uses, preferring instead to leave them for the pollinators. This is the perfect example of getting powerful aromas from very little plant material. We make a yeast starter jar (opposite page) with just a few blossoms. Be sure to use common broadleaf milkweed as the entire plant is edible; while you are not ingesting the blossoms, it is still best to avoid the narrow-leaf variety (*A. fascicularis*), as it is inedible.

Pines (*Pinus* spp.). All true pines are edible. Pines offer some nice piney flavor, headed toward turpentine with a citrus hint. We have used baby green cones for a wild yeast starter to inoculate apple juice to make cider. Caution: Norfolk Island pine (*Araucaria heterophylla*) and yew pine (*Podocarpus macrophyllus*), are not true pines, but both contain toxic compounds and should not be used at all.

Making a Wild Yeast Starter

Because yeast is everywhere, it can be harvested and reproduced to use in cider. We have done dozens of experiments, and as you might expect, we've had varying results. To be fair, we are adventurous eaters and drinkers, but we do have standards, and Kirsten is pretty picky about libations. Still, we've had way more successes than disasters.

A huge advantage to making a starter is that you make a small amount, so if it is not performing as you would like you are not out very much juice. This also gives you a chance to see how it responds to being fed more sugar, such as white sugar or honey. You can learn a lot by observing and tasting small batches that you make in pint jars. Another advantage is that you can propagate it when the ingredients are fresh, keep feeding it, and use it whenever you want — not just in a particular season. Lastly, you can isolate the strains you love to repeat — and that is the first step in a wild yeast becoming domesticated by you.

GATHERING WILD YEASTS FROM THE AIR

We collect the yeasts directly off our material of interest, but you can also make a wild yeast starter from the air in a specific part of your garden or orchard. Simply fill a sterilized jar three-quarters full of apple juice and cover with a piece of cheesecloth, secure it with a rubber band, and head out to your favorite spot, which is ideally among plants, bushes, or deciduous trees. We have found this works best in the spring and fall and in the evening, when the world is settling. Tuck the jar in a safe place under a bush or next to a tree, and let it sit undisturbed for 12 to 24 hours. Bring it in, cover it with a tight lid or place an airlock on the jar, then continue with the same process as making a starter jar.

To propagate the yeasts, you will put them in a sugary environment, where their population will increase to a point where you can pitch them into your cider. We use pasteurized apple juice to acclimatize the yeasts to the food source they will have when we pitch them in cider. You can either purchase pasteurized juice or pasteurize your own freshly pressed sweet cider by warming it to 160°F/71°C. Then pour it into a hot, sterilized jar, top with a lid, and allow to cool. You can also make a 20 percent sugar solution with boiling unchlorinated water and white sugar, though sometimes the sugar doesn't have enough "nutrients" for a strong start. The recipes that follow are for capturing other wild yeasts; for how to use the yeasts on your apples, see page 214 for making a wild yeast starter with fresh apple juice.

THE HARDINESS OF WILD YEAST

If you want to keep a batch of wild yeast active year-round, treat it like a sourdough starter. Remove some of the liquid to jumpstart a wild cider (page 141), if desired, and feed it pasteurized apple juice. If you keep it in the refrigerator, you will only need to feed it once a month or so.

We will come clean here. We have a lot to manage with two refrigerators of ferments, all the ciders, and the shelves of vinegar in varying stages — not to mention the wall of miso crocks. Things get forgotten and even "lost" behind other projects, especially pint jars of yeast starters. To our surprise, even after months of neglect at room temperature, sometimes these little jars of cider on their lees could still be woken up and used — that is the beauty (and hardiness) of spores. We are not going to tell you to not feed your yeast; we have lost plenty of starters from neglect too. But it can be possible to revive them.

Plum Blossom Cider

We make this lovely clean-tasting and bright spring cider with blossoms from our French prune plums. The instructions below are for a wild yeast starter with purchased juice, but we have also used plum blossoms to infuse already made ciders (follow the instructions for Wild Violet Cider page 250 for an infusion). Initial notes of bubble gum fade to floral with a lovely amaretto nose as the cider matures. This very light, pale cider still has some effervescence after it's been bottled and aged for 6 months, but if you want more sparkle, condition the cider at bottling with a bit of honey, which will give it a nice fizz and elevate the floral flavor. The recipe below is for putting the blossoms directly into your gallon of cider, but you could also make a plum blossom starter following the directions on pages 214–215.

Note: There is a lot of Internet kerfuffle about consuming plum blossoms due to their cyanide content. It is true that the leaves and seeds of plums (and less than trace amounts in blossoms) contain cyanogenic glycosides that can produce cyanide in your stomach, but you would have to eat a cup of seeds to have any negative effects. Given that the blossoms are filtered out, this shouldn't be a concern.

YIELD: 1 GALLON

- 1 gallon (3.8 L) apple juice
- 1 cup (80 g) loosely packed plum blossoms
- 1 (12-ounce/350 mL) bottle dry cider

1. Remove a little bit of the juice from the gallon glass jug, add the blossoms, and return as much of the juice as will fit, leaving about 2 inches before the cap.

2. Measure the SG of the juice with a hydrometer and record it in your cider log or on a piece of masking tape attached to the jug.

3. Insert the airlock and fill to the appropriate level with either fresh water or a neutral distilled spirit.

4. Let it sit in an environment where the temperature is between 55°F/13°C and 65°F/18°C. Wild yeasts like a cooler temperature, so the cooler in this range the better. Wild yeasts typically are slow to start so don't expect to see bubbles in your airlock for a couple of days or more.

5. The primary fermentation is finished when half to three-quarters of the sugars have been consumed, which you can determine by noting no bubbles being produced or if the SG is below 1.020. Depending on the wild yeasts present, the amount of sugar in the juice that they had to process, and the temperature, this can take as quickly as 2 weeks but probably more.

continued on next page

Plum Blossom Cider
continued

VARIATION:
ROSE PETAL CIDER

Every time we open a bottle of this cider, we are amazed at how delicious it is. Now we make it twice a year. In spring, we use the early blooms of the Gertrude Jekyll rose, which has a quintessential old-rose fragrance (a grandma smell for sure — you wouldn't think it is a good idea, but it actually is). In fall, we use a pinkish white climbing rose that grows on our trellis.

Use roses that are just beyond bud and early in their bloom. Unlike the recipe, you will pick off the petals instead of using the whole blossoms. Follow the recipe, using 80 grams of rose petals.

6. Sanitize a 1-gallon glass jug and siphon hose.

7. Using the siphon, rack your cider into the sanitized 1-gallon container, making sure to draw off all the cider above the lees, without drawing the lees out, which should be pretty minimal.

8. Add enough bottled cider to top off the racked cider to within a couple of inches from the top of the new 1-gallon glass jug to minimize air contact. Insert the airlock and fill to the appropriate level with either fresh water or a neutral distilled spirit.

9. Ferment in the same cool environment for at least 2 months.

10. Take one last SG measurement and calculate your final ABV. Taste, and if you would like it sweeter, back-sweeten it following the technique on page 106.

11. Siphon the cider into clean bottles, secure the tops, and store for at least 1 month before cracking one open.

Blossom and Petal Ciders

This is a basic recipe for making cider using a starter jar of yeast from full blossoms or just petals. Each blossom offers a unique take on the flavors in the cider. Although they are gorgeous, the flavor of cherry blossoms is pretty dull. Instead, look for basil blossoms or sunflowers (the latter yields a cider with the distinct nutty flavor of sunflower seeds). This method is a perfect way to make sure that you will like the flavor before committing to a large quantity of juice. Start this process at 1 to 2 weeks before you plan to make the cider.

YIELD: 1 GALLON

BLOSSOM AND PETAL YEAST STARTER JAR

1½ cups (355 mL) pasteurized apple juice

2–5 grams flower petals (such as basil blossoms or sunflowers)

CIDER

1 gallon (3.8 L) pasteurized apple juice

1 (12-ounce/350 mL) bottle dry cider

1. Prepare the starter jar with the petals and juice following the instructions on pages 214–15.

2. Taste the starter juice as it ferments and make sure that you like the flavors. As you monitor the ferment, you will see how it changes and continues to ferment the sugars.

3. Once the starter has finished fermenting, you can begin to make the cider.

4. Sanitize a 1-gallon glass jug and an airlock with a no-rinse sanitizer.

5. Shake up the starter jar to stir up the yeasts that have settled on the bottom of the jar. Pour the contents into the 1-gallon glass jug, making sure to strain out only the petals.

6. Add enough pasteurized juice to the glass jug to fill within 1 or 2 inches of the top. Measure the SG of the juice with your hydrometer.

7. Insert the airlock and fill to the appropriate level with either fresh water or a neutral distilled spirit.

8. Place your cider in an environment where the temperature is between 55°F/13°C and 65°F/18°C. Wild yeasts like a cooler temperature, so the cooler in this range the better. Wild yeasts typically are slow to start so don't expect to see bubbles in your airlock for a couple of days or more.

9. The primary fermentation is finished when half to three-quarters of the sugars have been consumed, which you can determine either by noting no bubbles being produced or by taking an SG reading. Note your initial SG reading to determine if it's reached the desired level of 1.020 or below. Depending on the wild yeasts

continued on next page

present, the amount of sugar in the juice that they had to process, and the temperature, this can take as quickly as 2 weeks but probably more.

10. Sanitize a 1-gallon glass jug and a siphon hose.

11. Using the siphon, rack your cider off into the sanitized jug, making sure to draw off all the cider above the lees, without drawing the lees out.

12. Add enough bottled cider to top off the racked cider to within a couple of inches from the top of the new glass jug to minimize air contact. Insert the airlock and fill to the appropriate level with either fresh water or a neutral distilled spirit.

13. Ferment in the same cool environment for 2 months.

14. Take one last SG measurement and calculate your final ABV. Taste, and if you would like it sweeter, back-sweeten it following the technique on page 106.

15. Siphon the cider into clean bottles, secure the tops, and store for at least 1 month before cracking one open. The bottles can be stored in a cool environment out of direct sunlight for 6 to 10 months, but at that point the cider will lose some of its sparkle. Note: Some of the wild yeast blossom ciders can turn dull or sour after 1 year.

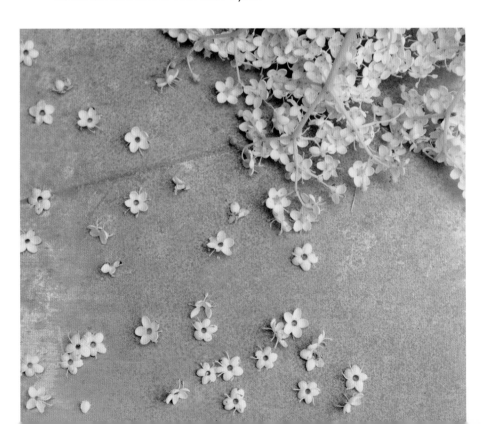

CHAPTER 7
INFUSED CIDERS

An infused cider is one in which botanicals, oak, spices, berries, flowers, or whatever else you dream up, have been added after primary fermentation and even after partial aging for the purpose of imparting flavor and/or an herb's healing energy. For example, fern licorice root stimulates the immune system and the bitter bark of the wild cherry or dandelion root will aid digestion. The cider is already made, so you are not trying to capture the yeasts, or any of the other microbes for that matter; you are only after the aromas and flavors of the addition. Here we share a few different processes of infusion for getting the most out of your botanicals. This information is by no means exhaustive. Instead, we hope it will encourage your creative side and motivate you to start looking around your region for inspiration.

Botanicals have a long history of being partnered with alcohol. These beverages were also considered medicine, and the ferment was a way to preserve the properties of the sacred plants. Scandinavians made beer with evergreen boughs and the Norwegians with juniper berries. Nettles, St. John's wort (that word right there, *wort*, is a clue to its early use), and lemon balm all were important ingredients in beer. The biodiversity of the ales and libations was vast due to all the botanicals used.

We recommend that you consult with a qualified healthcare practitioner before using herbal ingredients, particularly if you are pregnant, nursing, or on any medications.

HOW AND WHEN TO INFUSE YOUR CIDER

You can infuse any cider at any point in the process. Generally, we choose a time during the secondary fermentation. Often the botanicals that you add, especially flowers, will bring a little more sugar to the yeast and the ferment will wake back up. There is an advantage to adding the botanicals when you are racking because it minimizes oxygen exposure, and giving the yeast this extra bit of food enlivens the fermentation. We also add the botanicals to a cider that has already fully fermented; at this point you may not get as much fermentation on the added botanicals, but they will still impart some nice light flavors. We like to use infusions as a way to spice up dry ciders that are fine but unremarkable; the infused botanicals make them shine.

When you are ready to add your botanicals to a cider, remember, your goal is to minimize the amount of oxygen that is exposed to the cider. You

The cider on the left is a basic dry cider that was used to produce all three ciders on the right: Chai-der (page 253), nettle "leaf tea" cider, and Chocolate Cider (page 245).

have two choices. You can place the botanicals in the bottom of the new container (gallon glass jug or other fermenter) and use a racking cane to syphon the cider into the container. Or you can simply add them to the top of an existing container of cider. You can add your fresh or dried botanicals to the cider, or you can prepare the botanicals first by using one of the two following methods, both of which come from Sean Kelly (see profile on page 230) — making tea leaves (page 228), or perfuming botanicals with the smoke from resinous plants (page 233).

SPICED CIDER

According the United States Association of Cider Makers (USACM), spiced ciders are defined as "ciders made with any combination of spices, herbs and/or botanicals added either before or after fermentation." We have had a lot fun with this idea, and our ideas continue to expand as we find new herbs or flower blossoms with which to experiment. We know those of you who are adventurous will take some of the ideas presented and soar.

1. After your cider has gone through primary fermentation, place your botanicals in the bottom of a sanitized carboy.

2. Using a siphon, rack your cider onto the botanicals and seal the carboy with an airlock. Follow the recipe instructions for the secondary fermentation.

3. When your cider is ready to bottle, siphon the cider into clean bottles, making sure to draw off all the cider, leaving the lees and botanicals in the carboy.

Oxidizing Leaves for Botanical Teas

One way to prepare leaves for infusion into cider is to oxidize them in the same way that *Camellia sinensis* goes through variations of controlled oxidation to produce tea. Why do this? Because oxidation releases other compounds and flavors that can be utilized and fully realized in a cider. It also makes astringent leaves less so. Just to be clear: oxidation and fermentation are not the same thing. There are no microbes involved in oxidation. Rather, it is a chemical reaction involving oxygen (and enzymes), resulting in browning or discoloration. In the case of leaves, the enzymes cause the oxygen to be absorbed.

If oxidation goes too long, the good compounds are lost. There are two ways to halt oxidation. One is by removing all the oxygen, which is what we do when we attach an airlock to a ferment; we allow the carbon dioxide to push the oxygen out without letting in any new oxygen. The other is by denaturing, or "killing," the oxidation enzymes. This is done with a quick heat treatment like steaming or toasting.

Now that you have had a quick primer on oxidation, let's learn how to oxidize leaves into "tea botanicals." You can try this method with any number of edible garden-grown or wild-foraged leaves. As always, remember not to harvest along roadsides or areas that are sprayed with chemicals (see page 206 for foraging precautions). Start with young dandelion greens and nettle leaves in spring and move through the season, trying young rose leaves, raspberry leaves, and the like. Because the process includes steaming, these won't yield wild yeasts, as they will be killed by the heat.

Start with small batches to get a feel for the process. It is a little time consuming but also meditative in its way. It is also a chance to touch and sense what the plant has to offer. Through the process you can also taste the leaves as they change. This gives you a chance to get to know the qualities of the plant.

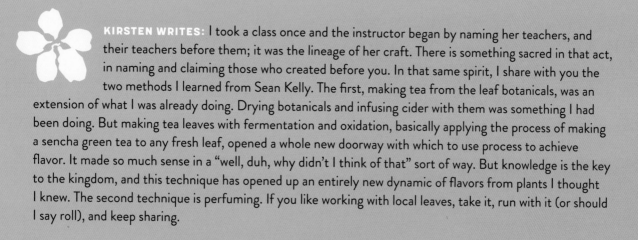

KIRSTEN WRITES: I took a class once and the instructor began by naming her teachers, and their teachers before them; it was the lineage of her craft. There is something sacred in that act, in naming and claiming those who created before you. In that same spirit, I share with you the two methods I learned from Sean Kelly. The first, making tea from the leaf botanicals, was an extension of what I was already doing. Drying botanicals and infusing cider with them was something I had been doing. But making tea leaves with fermentation and oxidation, basically applying the process of making a sencha green tea to any fresh leaf, opened a whole new doorway with which to use process to achieve flavor. It made so much sense in a "well, duh, why didn't I think of that" sort of way. But knowledge is the key to the kingdom, and this technique has opened up an entirely new dynamic of flavors from plants I thought I knew. The second technique is perfuming. If you like working with local leaves, take it, run with it (or should I say roll), and keep sharing.

Plan to process your leaves soon after picking. These leaves are still "breathing" and giving off heat; if left in a pile this heat will begin to break them down, but not in the controlled way you are working toward. Right after picking, leaves are traditionally fanned with humid air to keep them moist and fresh.

1. Pick young leaves (new growth is especially vibrant) of edible plants.

2. Put a sieve over a pot of simmering water. Place the leaves in the sieve and steam lightly, rotating them as they steam so that the steam works evenly through the leaves. It's a little counter-intuitive, but steaming releases water from the leaves. You are looking for roughly a 50 percent loss of moisture. The leaves will look a little thinner and wilty soft, but they will be bright green. You don't want them cooked or falling apart. Note: The longer they are steamed, the more astringency (in cider, some is good) and flavor will be lost.

3. Spread out the leaves on a clean surface to dissipate the steam and heat as quickly as possible. Note: If the leaves are kept too warm, they will begin to lose their aromas.

4. Press and roll the leaves until you see the color richen, then allow to rest for about 10 minutes. This step softens the leaves, releases more moisture, and breaks up the cell walls, exposing them to oxygen, which begins the oxidation of the leaves. Note: You don't want the leaves to dry out.

5. Repeat step 4, rolling and resting. Each time you will be reducing the moisture content evenly.

6. You have two choices for the final drying to further draw out the aromas. You can roll all the leaves in one direction to make little cigar bundles and dry them using a hair dryer or a dehydrator. Or you can quickly toast the leaves in a dry pan set over medium heat to remove the rest of the moisture content.

7. When dry, the leaves will still have a slight moisture content. You will know this because the leaves will be somewhat pliable and will stay fairly intact. They should not crumble to the touch. The leaves are now ready to infuse in a cider as described on page 226.

8. Store the dried leaves in an airtight jar. If you will be storing for the long term, consider vacuum sealing. The leaves will continue to oxidize over time, and once they are fully oxidized, they will taste stale.

Sean Kelly of WildCraft Cider Works

From the outside, Sean Kelly may appear to be an artist and flavor maker, but he will tell you that he is in the business of resource management. As you will discover, his approach is about bottling both the wild expression of the landscape circling WildCraft's cidery in Eugene and the pragmatism of using what is available.

Sean's cider is terroir specific, showcasing the flavor profiles in the 35-mile radius that defines his working area. His resources are fruit orchards whose trees, which are abandoned or no longer commercially viable in the conventional sense, still bear bushels and bushels of fruit. Other resources are the orchards that have long been plowed under and are now acres of wildly growing pear rootstock or seedlings. At the time of our conversation, Sean and his crews were picking and managing pome fruit from 240 private properties, and they had contracts to pick in 8,000 acres of public land. It is arduous, but as Sean points out, you can't buy these ingredients.

He is always on the lookout for potential ingredients, scouting the landscape, and gleaning its bounty.

For example, WildCraft's perry is made from pears that grow on pear rootstock that have naturalized on thousands of acres. The small fruit come in a rainbow of blushed colors and are high in tannins and acids and lower in sorbitol. He has found an orchard in the foothills of the Cascades that was planted by French fur trappers, as well as a cherry variety called the Black Republican that was developed in the area by abolitionists. Pineapple quince, medlars, persimmons, elderflowers (and the list goes on), with the yeasts they harbor, are all part of his palette.

These raw materials are practical and ecological. Sean doesn't see the rationale in trucking food (or beverages) all over the country, or the world for that matter, and real cider is food. He points out that we humans know better. He sees cider as the lubricant to stimulate important conversations about how we manage the land. He works to educate the public and thinks that local land stewards should be given a place to channel their resources to local businesses. In his view, cider is an opportunity for symbiosis between us and our landscape. In that vein, he developed a community cider (the first we know of) made from the fruit community members bring from their yards every fall to be upcycled into cider. They get fresh juice on the spot or a voucher for cider later.

WildCraft ciders also rely on wild apple yeasts. He doesn't make yeast starters from the many botanicals he uses. When you taste a WildCraft cider you are truly tasting the essence of the landscape, as well as getting the health aspects from the soil up through the complex web of inputs that grow the fruit. It is a matter of trust for Sean — he trusts the ingredients, the microbes, the development each of the parts will take, and the human senses to guide the process.

Sean doesn't control the temperature in the cidery, which gives his cider a certain seasonal variation and is closer to how it was traditionally made. He does, however, control the music. Microflora have an energy; some years the energy in the barrels is intense and in other years the energy is gentle.

By the same token, Sean has observed that the microbes also listen and respond to human energy; especially their response to the music that is playing in the cidery. He is quick to point out that he doesn't know if there is any science to this, but he's noticed that the more delicate fermentations with lower starting yeast colonies will be affected grossly by negative human energy and distortion sounds. This translates as stuck fermentations that will go acetic very rapidly. Vibrations such as steady bass or rhythmic keys and harmonies can stimulate either dominant yeasts or even change the direction of fermentation altogether to make a different strain dominant. To this end, ciders fermenting at WildCraft sometimes get their own beats in what Sean dubs the "tote bang sessions." Sometimes when bands come to play in the tasting room, afterward he invites the musicians to go into the cidery to bring up the energy by playing rhythmic beats on the fermenting intermediate bulk containers (IBCs). It is with this perceptive observation that Sean has evolved his business and production methods to produce truly unique ciders.

Perfuming

Perfuming is a technique for adding smoky botanical notes to cider. It is a good method for making a small amount of botanicals go a long, long, way; in other words, this technique offers powerful infused flavor. It is a way to capture resins — the saps used for medicine throughout time (see box, below) — or plants with resinous qualities (like hops and hemp). By smoking these plants and capturing the smoke within another botanical, you can imbue the cider with the qualities of the smoked plant. We often use chamomile or calendula flowers for this, as they soak up the smoke well, and while they do impart their own flavor, it is not overpowering. You can also think of perfuming as being similar to charring, a method used with things like rosemary and citrus to add flavor to food and cocktails, the difference being that here you are capturing the smoke.

WHAT ARE TREE RESINS?

Simply put, resin is the tree's sap that works as the tree's defense when it is damaged. The resin covers the wound where the bark has been compromised, thereby sealing off the tree from insects, parasites, and damaging microbes. When exposed to oxygen, resin hardens over time, creating a scab that allows the tree to continue to fend off harmful influences. When you are walking in the forest, you will often see these scabs as hardened orange or goldish drips. As you may suspect, the resin is part of the tree's immune system and has healing qualities (for the tree's own healing) as well as antifungal and antimicrobial properties to prevent decay. For this reason, humans have used these resins medicinally through the ages.

Resins are chemically complex and give off an assortment of aromas — bitter, spice, citrus, pine, balsam — that are rich and can be sweet, even somewhat floral. These qualities have made resins important and used as incense in spiritual ceremonies to bring cleansing and purification or protection, to name a few. If you have ever burned a single resin as incense, you may know that the feeling it brings to your senses and your body can be profound. When using resins, you should only use those that are known to be safe for human consumption like amber, balsam of Peru, and myrrh. Other resins are like those found in cannabinoids (see CBD Cider, page 241).

Tree resin

1. Heat a stainless steel pan or bowl or a cast-iron skillet over high heat until quite hot. Drop the resinous botanical on the hot pan, immediately turn down the heat to medium-low, and place a 1-quart jar, top down, over the botanical. (Note: If you are using something particularly pitchy, you can put it on a square of aluminum foil in the bowl or skillet.)

2. Keep the jar on the botanical for at least 10 minutes, or until the jar is densely filled with smoke. You are aiming for steady, slow heat and a botanical with a low smoke point for a more pronounced aromatic profile. More importantly, low-heat smoke doesn't blow off as quickly, which is important in step 4.

continued on next page

3. Make sure the jar is cool enough to the touch. If the heat was low enough, it should be fine. If it is hot, turn off the heat, take the bowl or pan off the stove, and wait until the jar is cool enough to handle. Carefully, but quickly so you don't lose much smoke, lift the jar off the pan and add ½ to 1 cup dry herbs. Secure the lid.

4. Shake the jar. No smoke should escape. You will see the smoke dissipate as the herbs capture the smoke and resins. After a few minutes of agitating the jar, you will notice that you can no longer see the hazy smoke. Set aside the jar and let it sit a few more minutes until it is fully cooled. You now have a smoky, resin-rich botanical ready for infusing into cider (follow the steps on page 226). Use immediately or leave the jar sealed and use within a week or two so as to not lose flavors. You can keep it at room temperature.

BOTANICALS FOR INFUSING

Blue spruce (*Picea pungens*). The young tips lend a mellow lemony citrus flavor.

Chamomile (*Chamaemelum nobile*) or, if you want to go wild-style, foraged pineapple weed (*Matricaria discoidea*). When dried, either of these small herbal blossoms make a wonderful medium for perfuming, described on page 232. They balance out smoky flavors with their gentle floral herbal notes. Note: We find that after 3 or 4 weeks of infusion, chamomile can impart some bitter flavors.

continued on next page

aging the pine flavor recedes and you taste a strong forest floor flavor, which is not the most pleasant. We have spoken with other brewers who experience this same change, and who also have not seen any mold during production or in the bottle. We believe there are compounds that develop these off/moldy flavors over time. The cider is delightful early in the process and for a few months once bottled. Therefore, we suggest you drink this cider young, soon after infusing.

Elderberry/elderflower (*Sambucus* spp.). The elder plant, steeped in lore and tradition, is truly an elder in the medicines of man. We have made elderberry wine and elderberry cider for many years; it is an easily obtained wild berry in our neck of the woods. Although the ratio can change with the seasons, we have found that using about 4 cups of mashed elderberries to 1 gallon of juice is just right. The stems and leaves are toxic, so be sure to strip the berries and flowers off the stems before infusing. Given that elderflower wine is a very traditional European wine and Kirsten's mother has made a delicious elderflower cordial, it was natural for us to try using elderflowers to start a wild cider from pasteurized juice. The finished cider had some faults, which we found over and over again when using high-summer flower yeasts, though it was still drinkable with a pleasant floral hint. We have yet to use it as an infusion, but we suspect, or rather hope, it will make a delicious cider.

Hibiscus/sorrel/roselle (*Hibiscus sabdariffa*). We have done a lot of experiments with hibiscus. It offers a beautiful red hue to the cider, and we love the refreshing sour flavors it imparts. We have used it as an infusion in both the primary and the secondary ferment. We found that about 50 to 60 grams of whole, dried flowers to 1 gallon works well. We haven't yet found the right ratio for the finely ground tea version (called

Dandelions (*Taraxacum officinale*). The roots are the strongest bittering agent of this plant. Don't use them freshly pulled from the earth, as you will introduce more bacteria than you are bargaining for. Wash and chop the fresh roots, blanch them in boiling water for 5 minutes, remove from the heat, and let steep for 1 hour. You can also use dried roots, available at health foods stores — boil and steep as you would with fresh roots. We use just a few grams of dried dandelion root to make about ½ cup of steeped tea, which we add to 1 gallon of juice. We also make wild yeast cider with dandelion blossoms (see notes on page 208).

Douglas fir (*Pseudotsuga menziesii*). This is not a true fir at all, but the soft tips of its new growth are edible, delicious, and have a fir-like flavor. The Douglas fir tip ciders we've made were delicious and reminiscent of retsina (a strong Greek wine) at bottling and during the first 2 to 3 months of aging. However, with continued

CBD Cider

We found the most interesting flavors came from perfuming (page 232) the buds of the hemp plant in mellowing chamomile blossom. Hibiscus was a close second. Perfuming certainly activates and pulls out the cannabinoids, but to what extent they land in the cider is something a lab would need to test. This can be made as a pitched or wild yeast cider.

YIELD: 1 GALLON

½ cup (118 mL) unchlorinated water

½ teaspoon champagne-style yeast (Lalvin EC-1118, Red Star Premier Curvée or Premier Blanc) (optional)

1 gallon (3.8 L) preservative-free apple juice for pitched cider, or sweet cider for wild

1 (12-ounce/350 mL) bottle dry cider

FOR PERFUMING

1–2 buds hemp, one medium bud will offer a suggestion of the smoky cannabis flavor, increase if you want to really know your cider is smoky

¾ cup (15–20 g) dried chamomile or hibiscus flowers, or other dried herb of your choice

1. Sanitize a 1-gallon glass jug and an airlock with a no-rinse sanitizer.

2. If pitching commercial yeast, heat the water to 104°F/40°C and pour into a quart canning jar. Sprinkle the yeast over the hot water, stir gently, and let it sit for 20 minutes. Stir again, measure the temperature of the yeast mixture, and write that down in your cider log or on a piece of masking tape attached to the 1-gallon jug.

3. Measure the SG of the apple juice with a hydrometer and the temperature with a thermometer and record them in your cider log or on a piece of masking tape attached to the jug. If making a wild cider, skip steps 4 and 5.

4. If the temperature of the yeast and the juice are within 18°F/10°C of each other, proceed to step 5. If not, add ½ cup of the juice to the yeast mixture, gently stir, and wait for 5 minutes. Measure the temperature and if it's within the 18°F/10°C range, move to the next step. If not, add 1 cup more of the apple juice to the yeast mixture, stir, and wait for 5 minutes.

5. Pour the yeast mixture into the glass jug. Add some juice to the yeast jar, swirl it around to get all of the yeast remaining in the jar, and pour it into the glass jug. Add enough juice to the glass jug to fill within 3 inches of the top. Cover the opening with plastic wrap.

6. Place the glass jug on a surface that is easy to clean or set it on a tray or pan. Let it sit in an environment where the temperature is between 55°F/13°C and 65°F/18°C. Bubbles will slowly form after a few days, then build. If the liquid froths out the top and down the sides (though 1-gallon batches seem a bit calmer than their 3-gallon siblings), simply clean the sides and tray/floor with a wet sponge. When the bubbles stay below the top of glass jug, apply the bung or cap, insert the airlock, and fill it to the appropriate level with either

continued on next page

CBD
Cider *continued*

fresh water or a neutral distilled spirit. Depending upon the yeasts, the temperature, and the juice, this can take 3 to 10 days or sometimes more.

7. The primary fermentation is finished when half to three-quarters of the sugars have been consumed, which you can determine by noting no bubbles being produced or by taking an SG reading. This stage should take between 2 and 3 weeks. If using SG, note your initial SG reading to determine if it's reached the desired level. For example, if the initial SG was 1.060, now it would be 1.030 to 1.015 or lower.

8. A few hours before you are ready to rack the cider, perfume the CBD buds using the chamomile as the substrate to capture the smoke according to the instructions on pages 233–34.

9. After shaking the tightly closed jar until you don't see smoke, set the jar aside for about 1 hour.

10. Meanwhile, sanitize another 1-gallon glass jug, a racking cane, and a siphoning hose.

11. Drop the smoked chamomile into the sanitized 1-gallon glass jug and, using a siphon, rack your cider off into the glass jug, making sure to draw off all the cider above the lees, without drawing the lees out.

12. Add enough bottled cider to top off the racked cider to within a couple of inches from the top of the new glass jug to minimize air contact. Insert the airlock and fill to the appropriate level either with fresh water or a neutral distilled spirit.

13. Ferment in the same cool environment for at least 1 month or up to 3 months.

14. Take one last SG measurement and calculate your final ABV. Taste, and if you would like it sweeter, back-sweeten it following the technique on page 106.

15. Siphon the cider into clean bottles, secure the tops, and store for at least 1 month before cracking one open. The bottles can be stored in a cool environment, out of direct sunlight, for a year or more, but at that point the cider will lose some of its sparkle.

WHITE SAGE CIDER

We love what smoky white sage (*Salvia apiana*) does to cider. We want to say it is like a smudge in a bottle, but that doesn't sound very tasty, does it? However, like smudging, there is something that feels ceremonial about drinking it. If you make it, you will see what we mean — it is a special occasion cider, not an everyday cider.

Follow the instructions for CBD Cider (page 241) but use 8 to 10 dried white sage leaves instead of the CBD buds. White sage is native to the southwestern United States. However, drought and human growth have limited the number of wild plants to be found. Pick only a few leaves responsibly, grow your own, or purchase them from a responsible source.

CBD AND CIDER: THE OIL AND WATER PROBLEM

The cannabis plant contains a number of different chemical compounds (cannabinoids), the most famous being cannabidiol (CBD) and tetrahydrocannabinol (THC). CBD is one of the nonpsychoactive chemical compounds; THC is the psychoactive component. Hemp is a member of the cannabis family, but unlike other strains of cannabis, it has not been bred to have high levels of THC, so it won't get you high. It does, however, have high levels of CBD, touted to have many health benefits, from heart and bone health to improving mood and pain management. We live in cannabis-growing country, so while we don't grow it, it is easy to come by. We tried using buds as a wild yeast starter and weren't impressed with the results — there was very little action on the fermentation. When we infused it, the flavors were too strong to consume, and besides, there is no real advantage to a standard infusion as the low alcohol content of cider does not extract the cannabinoids.

There are a number of ways to extract the compounds — carbon dioxide and ethanol alcohol being the most common in the industry, vegetable oils and butter being the most common on the home scale. Given that butter will make a good brownie but not a tasty cider (who wants a buttery oil slick on top of their drink?), you will want to use ethanol — and the higher the proof, the better. The easiest way to give your cider a little shot of CBD is to squirt a little tincture in it, but where is the fun in that? Also, the flavor may taste a little medicinal. You could extract a little CBD in apple brandy, which would give you some compounds and better flavor, but this is more of a mixed drink than a cider — unless you then turn that brandy into a pommeau (page 275). To be fair, oil and water don't mix, so putting an oil-based compound with very strong flavor in a drink is challenging. Big industry knows it will be a hit and is working on the issue, but it hasn't quite gotten past the flavors of barnyard and bong water. We found that the best way to get some CBD into our cider was to perfume some herbs with hemp smoke, and then infuse the cider with those perfumed herbs.

Is it legal? Well, it depends. The laws around cannabis (and its derivatives) and the laws around alcohol vary from state to state, and when you combine cannabis and alcohol, well there is even more to navigate. Unlike cannabis grown for THC, which is strictly forbidden by federal law, hemp grown within the parameters of part of a 2014 Farm Bill that authorized a state pilot program is. That means that only CBD from industrial hemp is allowed in alcoholic beverages.

KIRSTEN WRITES: For about a week, I was perfuming hemp buds to try to come up with a pleasant combination of smoke and herbs. The house stunk as you might imagine a cannabis café in Amsterdam might. We laughed, imagining what the kids would think if they were to walk in. When I was done, there were eight sample batches on the counter, which we added to our cider. Every few days we would stick a straw down through the mat of herbs that "sealed" the cider and taste. The cider that had been infused with hemp-perfumed chocolate nibs ended up tasting like a pot brownie cider, which was confusing to the palate (check, intriguing), but tasty . . . well, no. Our son's words: "Mom, people don't eat pot brownies for the flavor."

Chocolate Cider

We know it sounds all wrong and yet is all right. Because so much of what we taste comes from what we smell, this cider works. You get a cocoa bouquet that feels like you have the golden ticket and you are Charlie in the Chocolate Factory; you taste chocolate, but then it's cider.

This recipe is a straight infusion after fermentation. Make a base cider you like (sweet or dry) and add chocolate husks. Husks are the winnowed chaff that comes off the nibs; they can be obtained from local chocolate makers or online. We first tried making a chocolate cider with cocoa nibs, but even after making over a dozen batches with varying techniques, none of the ciders were amazing. We believe that the rich fats and bitter compounds that make chocolate so delicious muddied the cider. Then we remembered our neighbors at Apple Outlaw (page 186) had made a chocolate cider using husks, so we called them. Soon we had husks in hand, courtesy of Apple Outlaw, who had gotten them from a local chocolate maker.

For a fun variation, put 3 quarts of black cherry or raspberry juice or purée into the fermented cider with the husks.

YIELD: 3 GALLONS

---

3 gallons (11.4 L) fermented cider that is ready to rack

2 cups (175 g) cocoa husks

3 quarts (2.8 L) black cherry or raspberry juice or purée

1 (12-ounce/350 mL) bottle dry cider

---

1. Sanitize a 3-gallon carboy and an airlock with a no-rinse sanitizer.

2. Place the cocoa husks into the clean 3-gallon carboy. If adding a fruit juice, add it now.

3. Using a siphon, rack your cider off into the carboy with the husks, making sure to draw off all the cider above the lees without drawing out the lees.

4. Add enough bottled cider to top off the racked cider to within a couple of inches from the top of the new carboy to minimize air contact. Reapply the airlock.

5. Allow to infuse in a cool environment, between 55°F/13°C and 70°F/21°C, for 1 week.

6. Take one last SG measurement and calculate your final ABV. Taste some of the cider used to measure SG. If you would like it sweeter, back-sweeten it following the technique on page 106.

7. Siphon into clean bottles, secure the tops, and store for at least 1 month before cracking one open. This cider can be stored in a cool environment out of direct sunlight for 6 months or so.

Hopped Cider

If you have ever seen the hop vine (*Humulus lupulus*), it will come as no surprise that this resinous green flower is a member of the family Cannabaceae, whose illustrious members include hemp and THC-producing cannabis.

Hopped ciders are simply made by introducing hops, sometimes in pelleted form, into ciders during the secondary fermentation phase; a method called dry hopping. We have grown hops along the south side of our farmhouse for years to help reduce the summer heat on our A/C-less abode but never thought of putting them in cider until one fateful lunch when one of us ordered an Anthem Hops Cider by Wandering Aengus Ciderworks and loved it. That fall we stuffed the sticky, caterpillar-like catkins (hop cones) into the necks of 5-gallon glass carboys of cider and waited to see what happened. The hopped flavor was fantastic, but cleaning the carboys and fishing out all of those soaked and disintegrating hops was not. Now we use a hop bag that has a drawstring, which is handy because once you have the hops in the bag and it's cinched up you can squeeze the bag into the sanitized carboy and let it hang inside from the drawstring.

Hops impart a bitterness, distinct aromas, and other flavors like citrus and pine. The flavor intensity varies by hops variety and method of inclusion. For an extra citrus boost, add one crushed black lime (loomi; see page 249). This cider can be made with commercial yeasts or wild yeasts.

YIELD: 3 GALLONS

¼ cup (59 mL) unchlorinated water

1 teaspoon Lalvin ICV D47 yeast (optional)

3 gallons (11.4 L) fresh sweet cider

1 ounce (28 g) pelleted/dry cone hops, or about 5 ounces (142 g) fresh-off-the-vine hops

1 (12-ounce/350 mL) bottle dry cider

1. Sanitize your 3-gallon carboy and an airlock with a no-rinse sanitizer.

2. If pitching commercial yeast, heat the water to between 104°F/40°C and 108°F/42°C and pour into a quart canning jar. Sprinkle the yeast over the hot water, stir gently, and let it sit for 30 minutes to allow it to cool to room temperature.

3. Measure the SG of the apple juice with a hydrometer and record it in your cider log or on a piece of masking tape on the carboy.

4. Pour the sweet cider into carboy, allowing enough room for the yeast starter (if pitching yeast) and 3 inches of air space. If using wild yeasts, skip step 5.

5. Pour the yeast mixture into the carboy. Add some juice to the yeast jar, swirl it around to get all of the yeast remaining in the jar, and pour it into the carboy.

continued on next page

6. Apply the bung or lid, insert the airlock, and fill the airlock to the appropriate level with either fresh water or a neutral distilled spirit. If possible, place your cider in an environment where the temperature is between 55°F/13°C and 65°F/18°C.

7. The primary fermentation cycle is finished when you don't see bubbles in your airlock. Double-check by measuring the SG, which should be nearing 1.010 or below. Depending upon the temperature, this might take 10 days to 2 weeks, though it could go longer if the temperature is in the 50s or below.

8. Sanitize another 3-gallon carboy, racking cane, and siphoning hose.

9. Fill a fine-mesh straining bag with hops flowers. Using a siphon, rack your cider off into the carboy, making sure to draw off all the cider above the lees without drawing the lees out, letting it pour through the suspended hops in the bag.

10. Add enough bottled cider to top off the racked cider to within a couple of inches from the top of the new carboy to minimize air contact. Secure the hop bag drawstring to the neck of the carboy with a piece of tape and insert the airlock on the neck (and over the drawstring). Fill the airlock to the appropriate level with either fresh water or a neutral distilled spirit.

11. Ferment in the same cool environment for 2 weeks.

12. Draw out a sample with a wine thief and taste for the level of hops you are looking for. If it's not there yet, check back in a week.

13. Take one last SG measurement and calculate your final ABV. Remove the hops bag and compost the hops. The mesh bags should be cleaned right away and allowed to air-dry. Taste some of the cider used to measure SG.

14. Siphon the cider into clean bottles, secure the tops, and store for at least 1 month before cracking one open. The bottles can be stored in a cool environment out of direct sunlight for about a year and the hops should help to preserve the cider, though the taste may change during that time.

MAKING BLACK LIMES FOR FLAVORING CIDER

Black limes (loomi) are limes that have been boiled in salt water and then dried whole until their insides turn jet black and their interior sugars ferment slightly. They hale from Persia and are used in Middle Eastern cooking. Aromatic, with sweet fermented notes, these preserved limes can add complex citrusy flavors to cider.

To make black limes, mix together 2 tablespoons salt and 4 cups water in a large saucepan and bring to a boil over high heat. Using a spoon, carefully drop four to six limes into the water and boil for about 5 minutes. Transfer the limes to a bowl of ice water to cool. If it is hot and dry where you live, place the limes on a rack in the hot sun until dry, rotating once in a while, for 4 or 5 days. If that isn't an option, use a long needle and strong thread to thread the limes end to end and hang them horizontally in a sunny window until dry. You can also dry them in a food dehydrator, or an oven set to 150°F/65°C for 2 to 3 days. When properly dry, the outside will be anywhere from light greenish tan to dark brown or nearly black and the limes will feel quite light and hollow. When cracked open, they should have a shiny black flaky interior.

To use the black limes in cider, crush two of the limes and add them to the cider, or pulverize one lime with a blender, pour the powder into a tea bag, and drop it in the cider. Start tasting your cider after 1 week and keep tasting until you like it. We found 2 weeks gives the flavor of a bright summer evening — refreshing without being overwhelming. This cider reminds us of a Radlermass, the German summer cyclist's (radler) pint (mass) that is half beer and half citrus soda. We also pair loomi with our woodland ciders, and it is delicious in the hopped cider. The loomi cider ages well.

Wild Violet Cider

Violets are a small delicate flower, yet they stand up and get noticed in the bottle. The fragrance is beguiling. Your nose is the first to take in the violet notes followed by the pleasing floral flavor when it hits your tongue. We love this cider; it is especially fun to bring it out for guests as it is a surprising flavor.

We often make very plain dry table ciders for the purpose of infusing flavor from field and forest. In southern Oregon the cheery purple violets come around in February and offer the first taste of color and impending spring. See the box on the next page for notes on violets and harvesting. This recipe has you making the cider first (either pitched or wild-style), which you should start a few weeks before the violets are in bloom. But feel free to use a cider you have ready, perhaps one you made in the fall.

YIELD: 1 GALLON

¼ cup (59 mL) unchlorinated water

½ teaspoon Lalvin 71B or ICV D47 yeast (optional)

1 gallon (3.8 L) fresh sweet cider

½ cup (4–5 g) violet flowers (green sepals are fine)

1 (12-ounce/350 mL) bottle dry cider

---

1. Sanitize a 1-gallon glass jug and an airlock with a no-rinse sanitizer.

2. If pitching commercial yeast, heat the water to 104°F/40°C and pour into a quart canning jar. Sprinkle the yeast over the hot water, stir gently, and let it sit for 20 minutes. Stir again, measure the temperature of the yeast mixture, and record it in your cider log or on a piece of masking tape on the jug.

3. Measure the SG of the sweet cider with a hydrometer and the temperature with a thermometer and record them in your cider log or on the piece of masking tape attached to the jug. If making a wild yeast cider, skip steps 4 and 5.

4. If the temperature of the yeast and the juice are within 18°F/10°C of each other, proceed to step 5. If not, add ½ cup of the juice to the yeast mixture, gently stir, and wait for 5 minutes. Measure the temperature and if it's within the 18°F/10°C range, move to the next step. If not, add 1 cup more of the apple juice to the yeast mixture, stir, and wait for 5 minutes.

5. Pour the yeast mixture into the jug. Add some juice to the yeast jar, swirl it around to get all of the yeast remaining in the jar, and pour it into the jug.

6. Add enough juice to the jug to fill within 3 inches of the top. Cover the opening with plastic wrap. Insert the airlock and fill it to the appropriate level with either fresh water or a neutral distilled spirit.

7. Let it sit in an environment where the temperature is between 55°F/13°C and 65°F/18°C.

8. The primary fermentation is finished when half to three-quarters of the sugars have been consumed, which you can determine by noting no bubbles being produced or by taking an SG reading. Note your initial SG reading to determine if it's reached the desired level. For example, if the initial SG was 1.060, now it would be 1.030 to 1.015 or lower. Taste the cider you used to measure the SG and write any tasting notes in your cider log or on a piece of tape attached to the jug. Depending upon the yeasts, the temperature, and the juice, this can take 10 days to 2 weeks, or sometimes more if the temperature is in the 50s°F/10s°C or below.

9. Sanitize another 1-gallon glass jug, a racking cane, and a siphoning hose.

10. Drop the violet flowers into the sanitized jug and, using a siphon, rack your cider off into the jug, making sure to draw off all the cider above the lees, without drawing the lees out.

11. Add enough bottled cider to top off the racked cider to within a couple of inches from the top of the new jug to minimize air contact. Insert the airlock and fill to the appropriate level with either fresh water or a neutral distilled spirit.

12. Ferment in the same cool environment for at least 1 month, or up to 3 months.

13. Take one last SG measurement and calculate your final ABV. Taste, and if you would like it sweeter, back-sweeten it following the technique on page 106.

14. Siphon the cider into clean bottles, secure the tops, and store for at least 1 month before cracking one open. The bottles can be stored in a cool environment out of direct sunlight for a year or more, but at that point the cider will lose some of its sparkle.

WILD VIOLETS

Pansies, Johnny-jump-ups, bedding violas, and sweet violets are all classified as violas, but they are different varieties. They are native to the temperate climates in the Northern Hemisphere but are also found in Australia and the Andes Mountains. Depending on where you live, your wild violets may be a different variety than ours. We have not had experience with wild yellow violets, and we read somewhere that they can taste soapy. A little research revealed that the fragrance in all violets is from the compound ionone, and how you experience ionone has to do with your genes — some people experience the lovely, classic violet aroma and taste, others taste soap, and some can't smell or taste violets at all. (African violets are different altogether.) If you enjoy violets, then this is a fun cider ingredient. You can also make a wild yeast starter and pitch in purchased juice in the spring.

Blue violets are one of the first flowers to appear in lawns and shady edges of the forest. They are sweet and have a distinctive floral, old-fashioned aroma as they harken back to a time gone by. The characteristic violet perfume imbues the cider, making a cider that tastes wonderful. If you can taste ionone, this is a cider to try.

Chai-der

Apples, pumpkin pie spice, and fall go hand in hand. Chai tea fits right in — its spices of cinnamon, cardamom, and ginger, to name a few, go very well with apples. This is an easy infusion that creates a bright, lightly spiced cider with a hint of honey. This recipe adds honey to bottle condition and back-sweeten, but of course you may omit this step if you prefer. The black tea adds some tannins to the apple juice. Use your favorite loose-leaf chai tea mix. The one we use is made with Assam black tea, cardamom, ginger, fennel, cinnamon, peppercorns, and cloves.

YIELD: 1 GALLON

¼ cup (59 mL) unchlorinated water

½ teaspoon champagne-style yeast (Lalvin EC-1118, Red Star Premier Curvée or Premier Blanc)

1 gallon (3.8 L) preservative-free apple juice

20 grams loose-leaf chai tea mix

1 (12-ounce/350 mL) bottle dry cider

Honey for bottling (optional)

1. Sanitize a 1-gallon glass jug and an airlock with a no-rinse sanitizer.

2. Heat the water to 104°F/40°C and pour into a quart canning jar. Sprinkle the yeast over the hot water, stir gently, and let it sit for 20 minutes. Stir again, measure the temperature of the yeast mixture, and write that down in your cider log or on a piece of masking tape attached to the jug.

3. Measure the SG of the apple juice with a hydrometer and the temperature with a thermometer and record them in your cider log or on the piece of masking tape attached to the jug.

4. Pour the yeast mixture into the jug. Add some juice to the yeast jar, swirl it around to get all of the yeast remaining in the jar, and pour it into the jug. Add enough juice to the jug to fill within 3 inches of the top.

5. Insert the airlock and fill it to the appropriate level with either fresh water or a neutral distilled spirit.

6. Let it sit in an environment where the temperature is between 55°F/13°C and 65°F/18°C.

7. The primary fermentation is finished when half to three-quarters of the sugars have been consumed, which you can determine by noting no bubbles being produced or by taking an SG reading. Note your initial SG reading to determine if it's reached the desired level. For example, if the initial SG was 1.060, now it would be 1.030 to 1.015 or lower. Taste the cider you used to measure the SG and write any tasting notes in your cider log or on a piece of tape attached to the jug. Depending upon the yeasts, the temperature, and the juice, this can take 10 days to 2 weeks, or sometimes more if the temperature is in the 50s°F/10s°C or below.

continued on next page

Chai-der
continued

8. Sanitize another 1-gallon glass jug, a racking cane, and a siphoning hose.

9. Transfer the chai tea mix to the sanitized jug (or put it in a cloth infusion bag and drop it in). Using a siphon, rack your cider off into the jug, making sure to draw off all the cider above the lees, without drawing the lees out.

10. Add enough bottled cider to top off the racked cider to within a couple of inches from the top of the new jug to minimize air contact. Insert the airlock and fill to the appropriate level with either fresh water or a neutral distilled spirit.

11. Ferment in the same cool environment for at least 9 months to ensure that all the yeasts have run out of food and are inactive.

12. Take one last SG measurement, which should be below 1.000, and calculate your final ABV. Use honey to back-sweeten, if desired, by mixing together equal amounts of honey and hot water and stirring to incorporate. Add 1 teaspoon of this simple syrup to each clean bottle. Because of the long secondary ferment, the sweetness should remain.

13. Siphon the cider into the clean bottles, secure the tops, and store for at least 1 month before cracking one open. The bottles can be stored in a cool environment out of direct sunlight for a year or more, but at that point the cider will lose some of its sparkle.

Quick Fruity Cider

Here is an example of infusing cider with commercial fruit juice or fruit juice concentrates. Use juice that is only the selected fruit juice with no other juice in it, for example pure pomegranate, pure sour cherry, or pure pineapple. This recipe uses a fully fermented dry cider, but if you want some sparkle, use a cider that has some living yeasts and treat as if adding priming sugar (see page 116).

YIELD: 1 GALLON

1 cup (237 mL) pure single fruit juice concentrate or 2 cups (475 mL) pure single fruit juice

3½–3¾ quarts (3.3–3.5 L) finished cider

1. Sanitize a 1-gallon glass jug and an airlock with a no-rinse sanitizer.

2. Add the juice concentrate or juice to the jug.

3. Add enough dry, finished cider to the jug to fill within 2 or 3 inches of the top. Apply the bung or lid, insert the airlock, and fill to the appropriate level with either fresh water or a neutral distilled spirit.

4. Allow to infuse for 6 to 7 days or until you find the perfect spot, tasting after 3 days and again after 6. Note that if there are some residual yeasts in the finished cider, the cider will continue to ferment. Otherwise, you are introducing flavors and with time the flavors will come together. If fermentation does take place, allow it to finish before bottling.

5. Using a siphon, rack your cider off into clean bottles and secure the tops. This cider should be consumed within 3 or 4 months of bottling. Store in a cool environment out of direct sunlight.

Two cups pure cherry juice ready to infuse cider

Strawberry Cider

Close your eyes and think of a bowl of fresh strawberries just in from the field. Can you smell them? Strawberries are the most commonly consumed berry fruit in the world, and much of their popularity comes from their aroma, which is a complex mix of over 360 different volatile compounds.[19] Among these are esters, which are the most abundant compounds and responsible for the floral-fruit odors that we love. Strawberries also contain sulfur compounds in lesser quantities (thankfully). Most of these sulfur compounds increase during the final ripening and during biosynthesis and although they are small in number, they can be mighty to the nose, at least for some people (Kirsten thinks it smells like burnt rubber). We have found through experimentation that this sulfur smell can be reduced or mitigated altogether by infusing the finished cider with strawberries instead of introducing them during primary fermentation.

Frozen strawberries have a rich jammy aroma and flavor because the fruit's structure is destroyed by freezing, freeing up the esters in the phenols. Freezing also helps retain the color. This cider is cloudy, or as beer lovers call it, hazy.

YIELD: 1 GALLON

1 pound (450 g) frozen strawberries

3½ quarts (3.3 L) finished cider

1. Sanitize a 1-gallon glass jug and an airlock with a no-rinse sanitizer.

2. Thaw the strawberries and purée them in a blender or food processor.

3. Add the strawberry purée to the sanitized jug. Add cider to fill within 2 or 3 inches of the top. Apply the bung or lid, insert the airlock, and fill to the appropriate level with either fresh water or a neutral distilled spirit.

4. Allow to infuse for 6 to 7 days or until you find the perfect spot, tasting after 3 days and again after 6.

5. Using a siphon, rack your cider off into clean bottles and secure the tops. This cider should be consumed within 3 or 4 months of bottling. Store in a cool environment out of direct sunlight.

FRUIT INFUSIONS

There are many reasons why you may want to infuse a bit of fruit flavor into your cider. It could be that you want to add some tropical flavors, like pineapple, mango, guava, or passion fruit, or you want to add some more common fruits that aren't in season when the apples are ripe. In these cases, it makes more sense to purchase juice or frozen fruit (or freeze your own) than to use fresh fruit. Two other practical reasons to use juice or frozen fruit purée to infuse a cider are time and simplicity. You can infuse a finished cider with fruit whenever you want to, and in about a week it will be ready to drink. And both a purée and a juice will give the yeast more sugar to work with but won't cause a frothy spillover. If you infuse a finished cider with juice, the infused cider will also be clear —no need to wait for it to clarify.

The final alcohol levels of a cider are determined early in the process by the amount of natural sugars available to the yeasts to become alcohol. Apples, even very sweet dessert apples, have their sugary limits. To create a drink with an alcohol level more typical of wine (around 12 percent ABV), we need to provide the yeasts with more food.

In this chapter, we explore going beyond how juice naturally ferments, to create some truly powerful, yet tasty, ciders. One way is to concentrate the sugars that are naturally in the juice. For this we look to fire and ice. Sometimes you want something more — a slow sipper that distills the essence of the apple into something raw, simple, and bare. While distilling in your own home is still illegal in the United States, we include a recipe for pommeau that should be acceptable to local authorities, since you are simply mixing two legal beverages.

THE SWEET FREEZE

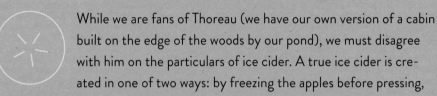

While we are fans of Thoreau (we have our own version of a cabin built on the edge of the woods by our pond), we must disagree with him on the particulars of ice cider. A true ice cider is created in one of two ways: by freezing the apples before pressing, or by freezing the juice and then drawing off the concentrated juice as it thaws and fermenting that. The most common way is to press your juice, freeze it, then thaw it slowly in order to extract the sugary liquid. This way, you are leveraging the fact that water and sugar have different freezing points. Water freezes first, causing the sugar to separate from the water. Upon thawing, the first part to melt is the ice crystals that contain most of the sugar. If you freeze the apples whole, press when partially thawed. The principle is the same. To be legally sold as "ice cider," commercial producers must freeze the apples or juice. But you have no such limitations and you can play around with this one.

If you decide to freeze freshly pressed apple juice, freeze it in a reused plastic gallon container. Pour off the juice as it melts, leaving some of the water that is last to melt, then ferment the concentrated juice after it comes to room temperature. This might sound like applejack, and it's close, but ice cider starts with fresh juice, while applejack starts with already fermented cider.

The biggest challenge to making ice cider is keeping the fermentation under control. You must stop it in time to hold on to that residual sugar. The yeasts are at the party and it's a good one; the sugar is flowing and there is a lot of it. It becomes a dance to find the right temperature — warm enough to keep them working then cold enough to shut them down.

Why go to all this work? Because you get a more intense apple flavor and aroma and a deeper and richer color, and it is a great way to use culinary apples, though, like keeving, the best apples are from old trees growing in low-nitrogen soils. Like ice wines, ice ciders are usually finished before all the abundant sugars are converted to alcohol, giving you a dessertlike drink that's better for sipping after a meal than chugging on a hot day.

Ice Cider

This cider is a good use of sweet eating apples because you get all that sugar with less of the other flavor elements of bitter, sharp, and sour. You will start with 3 to 4 gallons of juice, depending on how far you are hoping to reduce the liquid. Ideally you will end up with a little less than a gallon of concentrated juice to begin fermenting.

We add commercial yeast in this recipe because we want to make sure there is enough yeast to get the fermentation going. Presumably all the wild yeasts don't die in the freezer, but we don't know if they are always strong enough for consistency in this recipe. You can use a purchased yeast or make a wild starter with a quart of apple juice you set aside while the rest is freezing. Use it to start fermentation when the cider has been frozen and the thicker juice has been poured off. A wild yeast starter may not hit the higher alcohol levels, but it can be advantageous on the back end, when you want to stop the fermentation short.

To slow and stop the fermentation, some folks use sulfur dioxide or they pasteurize the cider. Instead, we keep it slow and cold and use racking as needed to "starve" the yeasts.

YIELD: 1 GALLON

- 3–4 gallons (11–15 L) unpasteurized sweet cider
- ½ cup (118 mL) unchlorinated water (not needed with wild yeast starter)
- ½ teaspoon Lalvin 71B yeast (or a wild yeast starter that has not been frozen)
- 1 (12-ounce/350 mL) bottle dry cider

1. Pour the juice into clean 1-gallon containers such as recycled plastic water jugs or food-grade buckets and freeze until solid. Using buckets can be easier because when it is time to pour the juice off, the chunk of ice won't block the spout.

2. To thaw, place the containers of frozen juice into a cooler for up to 7 days (this number is completely dependent on how slow your setup can keep the process). The slower the juice thaws, the more sugar you can pour off. For this reason, keeping it at refrigerator temperature (somewhere less than 42°F/6°C) is ideal but not necessary. The cooler method does a pretty good job, but the temperature inside gets warmer as the ice melts. Alternately, if you have a cool space, somewhere in the low 50s°F/10s°C, you can punch a small hole in the lid of the bucket or cap of the jug, and place the container upside down over a clean bucket to catch the juice as it thaws.

3. Sanitize a glass jug and an airlock with a no-rinse sanitizer.

4. Measure the SG of the juice with a hydrometer and record it in your cider log or on a piece of masking tape attached to the glass jug. For this cider the starting SG should be 1.125 or more. If it is less, it will be lower in alcohol

continued on next page

Ice Cider

continued

with less sweetness. If it is significantly lower, go through the freezing process again to draw out more water. Let the juice warm up until it is at least 60°F/16°C.

5. Heat the water to 104°F/40°C and pour into a quart canning jar. Sprinkle the yeast over the hot water, stir gently, and let it sit for 20 minutes. Stir again, measure the temperature of the yeast mixture, and write that down. If using a wild yeast starter culture, pour at least ½ cup of your active starter into the jug, using any extra to fill within 2 to 3 inches of the top. Measure the temperature and write that down in your cider log or on a piece of masking tape attached to the jug.

6. If the temperature of the yeast and the juice are within 18°F/10°C of each other, proceed to step 8. If not, add ½ cup of the juice to the yeast mixture, gently stir, and wait for 5 minutes. Measure the temperature and if it's within the 18°F/10°C range, move to the next step. If not, add 1 cup more of the apple juice to the yeast mixture, stir, and wait for 5 minutes.

7. Pour the yeast mixture into the jug. Add some juice to the yeast jar, swirl it around to get all of the yeast remaining in the jar, and pour it into the jug. Add enough juice to the jug to fill within 2 to 3 inches of the top.

8. Insert the airlock and fill it to the appropriate level with either fresh water or a neutral distilled spirit.

9. Let it sit in an environment where the temperature is between 55°F/13°C and 65°F/18°C for the first day or so, until you see signs of fermentation.

10. Lower the temperature to between 45°F/7°C and 50°F/10°C and watch the fermentation. If it speeds up, lower the temperature even more. We often use the refrigerator. A refrigerator dedicated to this can be adjusted, but you will often get that slow fermentation even in your regular fridge.

11. You can also rack the cider to reduce the nutrients, which will also slow down the fermentation. Draw off all the cider above the lees, without drawing the lees out.

12. Add enough bottled cider to top off the racked cider to within a couple of inches from the top of the new glass jug to minimize air contact. Insert the airlock and fill to the appropriate level with either fresh water or a neutral distilled spirit.

13. This may need to happen a few times before the primary fermentation is over. Racking will also keep the yeasts from autolyzing (a positive in some styles, unwanted in this one). This primary fermentation should take about 2 months.

continued on page 264

Ice Cider
continued

14. Measure the SG as the fermentation slows. The goal is for the SG reading to be 1.100 to 1.050 (no lower) for the finished cider. As you approach these numbers, begin the process to fully stop the fermentation. Your best tools to starve the yeasts are by racking or cold crashing by freezing (see next step).

15. If you have hit your target numbers and the fermentation is still going, rack your cider into a plastic 1-gallon jug, secure the lid, and put it in the freezer for a month or more. For home-based projects this works well. In Quebec, makers will freeze the cider for up to 9 months to stop the fermentation.

16. Check for the cider's stability because it still contains quite a bit of unfermented sugar — and if it is still fermenting and you bottle it, that can cause explosions. Test the stability by bringing the cider to a warmer temperature (or thawing it if it is frozen) and taking an SG reading. Let the cider sit for 24 hours at room temperature and observe the airlock for any bubbles. Take another SG reading and if the reading is the same, you are ready to bottle. If not, return the cider to the freezer for another week before testing it again.

17. For more insurance, bottle in Grolsch-style bale bottles, which have a little more give for pressure to release before exploding. After bottling, keep all but one in a cool spot. Test the fermentation by keeping one bottle warm for a month, then opening it to test for pressure.

FIRE

Fire cider (not to be confused with the herbal vinegar of the same name) came about in 1994 when Normand Lamontagne, a cidermaker, decided to concentrate the sugars in fresh apple juice by using a maple syrup evaporator and then fermenting the resulting juice. The finished product was a high alcohol (around 15 percent ABV) dessert cider. Simmered cider has a richer sweetness than ice cider, due to the effects of the heat.

To make fire cider at home, reduce your juice by simmering in a wide pot (wide and shallow is best, as a larger surface area will speed up the process) over medium to low heat, lid off. Don't let your cider boil. This needs to be slow and steady. When it boils, the sugars in the juice begin to go through the stages of candying, the first of which is syrup. You may want syrup if you are making a New England-style "boiled cider" or cider syrup, but it will cause difficulty in the fermentation. As the juice is reducing, check the SG — it should be 1.125 or more. From there, treat it as you would the Ice Cider (page 261). Since this juice is already heat-treated, you can also pasteurize it to stop the fermentation.

WHEN CIDER BECOMES WINE

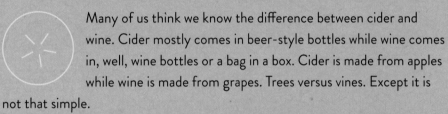

Many of us think we know the difference between cider and wine. Cider mostly comes in beer-style bottles while wine comes in, well, wine bottles or a bag in a box. Cider is made from apples while wine is made from grapes. Trees versus vines. Except it is not that simple.

Wine is a fermented drink that can be made from any fruit, including apples. Grapes just happen to be an excellent fruit for the task and have been for thousands of years. In the United States, if a cider passes from an ABV of 8.5 percent to 8.6 percent, it magically becomes classified as a wine. That's it — just a tenth of 1 percent and you cross the threshold from a cider to a fruit wine. We know, it seems pretty trivial, doesn't it? But as with all lines, you have to put it somewhere, don't you? It's also a line that Christopher finds himself willingly crossing to explore different flavor profiles when different types of sugar are added.

Apple Wine

When you add sugar to apple juice, you boost the amount of alcohol that will be made by the yeast. The type of sugar you use and how much of it is directly affected by your target final ABV and the taste you are going for. In this recipe we wanted something around 13 percent ABV. We started with store-bought organic apple juice with an SG of 1.060, which would have resulted in an 8 percent ABV. By adding 2 cups of sugar per gallon of juice, we raised the total SG to 1.098, which would result in an ABV of roughly 13 percent. Use this basic formula to adjust the quantity of sugar if you are looking for a different target ABV. You can, of course, substitute freshly pressed apples — just be aware that the sugar level changes as the SG changes.

Different types of sugars will affect the final flavor, especially when used in larger quantities. We suggest you start with a high-quality cane sugar. After you have made a batch with that, experiment with other types of sugar. We have used turbinado (also known as demerara sugar or raw cane sugar), corn sugar, brown rice syrup, and apple juice concentrate. These impart their own flavors to the mix.

YIELD: 3 GALLONS

½ cup (118 mL) unchlorinated water

1 teaspoon Lalvin ICV D47 yeast

6 cups (1.2 kg) granulated cane sugar

3 gallons (11.4 L) preservative-free apple juice

1 (12-ounce/350 mL) bottle dry cider

---

1. Sanitize a 3-gallon carboy, 5-gallon food-grade bucket, funnel, stainless steel whisk, and an airlock with a no-rinse sanitizer.

2. Heat the water to 104°F/40°C and pour into a quart canning jar. Sprinkle the yeast over the hot water, stir gently, and let it sit for 20 minutes. Stir again, measure the temperature of the yeast mixture, and write that down in your cider log or on a piece of masking tape attached to the 5-gallon bucket.

3. Mix together the sugar and the apple juice in the 5-gallon bucket and stir with the whisk until the sugar is dissolved. Measure the SG of the juice-sugar mixture with a hydrometer and the temperature with a thermometer and record them in your cider log or on a piece of masking tape attached to the bucket.

4. If the temperature of the yeast and the juice are within 18°F/10°C of each other, proceed to step 5. If not, add ½ cup of the juice to the yeast mixture, gently stir, and wait for 5 minutes. Measure the temperature and if it's within the 18°F/10°C range, move to the next step. If not, add 1 cup more of the juice to the yeast mixture, stir, and wait for 5 minutes.

continued on next page

APFELWEIN

German-style cider, or Apfelwein, uses culinary apples instead of cider apple varieties and has a classic winelike character. Its production, and most of its consumption, is concentrated in four regions of Germany: Hesse, Bavaria, Rhineland-Palatinate, and Baden-Württemberg.

Of the four, Hesse, which is the region including Frankfurt, is the largest producer and consumer of Apfelwein by far. In fact, if you go on a pilgrimage to find traditional Apfelwein, you would be well advised to start in this area, as the Hessians drink 10 times as much as the rest of Germany. You can expect the Apfelwein you find to be in the 5 to 7 percent ABV range, which clearly puts it in the cider range and not really what we think of as wine in terms of alcohol content.

Apple Wine *continued*

5. Pour the yeast mixture into the 3-gallon carboy. Add some juice to the yeast jar, swirl it around to get all of the yeast remaining in the jar, and pour it into the carboy. Add enough of the juice-sugar mixture to the carboy to fill within 3 or 4 inches of the top. Cover the opening with plastic wrap.

6. Place the carboy on a surface that is easy to clean or set it on a tray or pan. Let it sit in an environment where the temperature is between 55°F/13°C and 65°F/18°C. Bubbles will slowly form after a few days, then build. If the liquid froths out the top and down the sides, simply clean the sides and tray/floor with a wet sponge. When the bubbles stay below the top of carboy, apply the bung or lid and the airlock. Depending upon the yeasts, the temperature, and the juice, this can take 10 days to 3 weeks, given the higher sugars.

7. The primary fermentation is finished when three-quarters of the sugars have been consumed, which you can determine either by noting no bubbles being produced or by taking an SG reading. Note your initial SG reading to determine if it's reached the desired level. For example, if the initial SG was 1.100, now it would be 1.025 to 1.015 or lower. Taste the cider used to measure the SG and write any tasting notes in your cider log or on a piece of tape attached to the carboy.

8. Using a siphon, rack your cider off into another sanitized 3-gallon carboy, making sure to draw off all the cider above the lees, without drawing the lees out.

9. Add enough bottled cider to top off the racked cider to within a couple of inches from the top of the new carboy to minimize air contact. Reapply the airlock and fill it to the appropriate level with either fresh water or a neutral distilled spirit.

10. Ferment in the same cool environment for 4 months.

11. Take one last SG measurement and calculate your final ABV.

12. Siphon the cider into clean bottles, secure the tops, and store for at least 5 months to allow the higher alcohols to mellow a bit before enjoying.

Honey Cider (Cyser)

Cyser is a type of mead that is made with apple juice and honey. Mead is not that different from ciders and apple wines. Here we focus on bringing out those apple esters from the mead's balance of sweet/acid/tannin/alcohol. If possible, your juice should be an equal mix of tannic cider apples and acidic culinary apples. This recipe will make a light-bodied, clear, semisweet to dry cider similar to a Syrah or sauvignon blanc.

Honey ages well and brings in an extra element, partly because of the higher percentage of alcohol. Mead makers often joke that if you don't like your mead, put it back and check it again next year. Interestingly, the longer it ages, the more the honey comes forward in the flavor.

YIELD: 3 GALLONS

2 cups (475 mL) unchlorinated water

1 quart (0.9 L) quality honey

3 gallons (11.4 L) unpasteurized apple juice

1 teaspoon Lalvin 71B or ICV D47 yeast

1 (12-ounce/350 mL) bottle dry cider

1. Sanitize a 3-gallon carboy and an airlock with a no-rinse sanitizer.

2. Heat 1½ cups of the water until just boiling. In a 2-quart container, mix together the honey and hot water and stir until the honey is dissolved. Pour the honey mixture into the 3-gallon carboy and add enough apple juice to fill within 6 inches of the top of the carboy. Stir to combine the juice and honey mixture.

3. Measure the SG of the juice with a hydrometer and the temperature with a thermometer and record them in your cider log or on a piece of masking tape attached to the 3-gallon carboy.

4. Heat the remaining ½ cup water to 104°F/40°C and pour into a quart canning jar. Sprinkle the yeast over the hot water, stir gently, and let it sit for 20 minutes. Stir again, measure the temperature of the yeast mixture, and write that down in your cider log or on a piece of masking tape attached to the 3-gallon carboy.

5. If the temperature of the yeast and the honeyed juice are within 18°F/10°C of each other, proceed to step 6. If not, add ½ cup of the juice to the yeast mixture, gently stir, and wait for 5 minutes. Measure the temperature and if it's within the 18°F/10°C range, move to the next step. If not, add 1 cup more of juice, stir, and wait for 5 minutes.

continued on next page

Honey Cider (Cyser)
continued

6. Pour the yeast mixture into the carboy. Add some juice to the yeast jar, swirl it around to get all of the yeast remaining in the jar, and pour it into the carboy. Add enough juice to the carboy to fill within 3 inches of the top.

7. Insert the airlock and fill it to the appropriate level with either fresh water or a neutral distilled spirit.

8. Let it sit in an environment where the temperature is between 55°F/13°C and 65°F/18°C.

9. The primary fermentation is finished when half to three-quarters of the sugars have been consumed, which you can determine by noting no bubbles being produced or by taking an SG reading. Note your initial SG reading to determine if it's reached the desired level. For example, if the initial SG was 1.060, now it would be 1.030 to 1.015 or lower. Taste the cider used to measure the SG and write any tasting notes in your cider log or on a piece of tape attached to the carboy.

10. Using a siphon, rack your cider off into another sanitized 3-gallon carboy, making sure to draw off all the cider above the lees, without drawing the lees out.

11. Add enough bottled cider to top off the racked cider to within a couple of inches from the top of the new carboy to minimize air contact. Insert the airlock and fill to the appropriate level with either fresh water or a neutral distilled spirit.

12. Ferment in the same cool environment for 3 months.

13. Take one last SG measurement and calculate your final ABV. Taste, and if you would like it sweeter, back-sweeten it following the technique on page 106.

14. Siphon the cider into clean bottles, secure the tops, and store for at least 6 months. If you can, try to tuck these away for 1 year or more, as the flavor only gets better at that point.

FINDING
THE SPIRIT

There are two fundamental ways to concentrate the natural alcohols in cider: distillation (boiling the cider and capturing the condensation) and fractional crystallization (freezing the cider and capturing the melt). They are similar in two regards. First, they both work by taking advantage of the different properties of alcohol and water. In the case of distillation, it's the difference in boiling points between alcohol and water. In the case of fractional crystallization, it's the difference in freezing points. In both cases we continually separate alcohol from water. The second similarity is that they are both illegal in the United States without permits and blessing from the Department of the Treasury's Alcohol, Tobacco, Tax, and Trade Bureau (TTB).

Distilling requires the use of a still, and the U.S. federal government only permits people to own one at home for distilling water or essential oils. Not unlike cannabis, while it is illegal at the federal level, it is legal in some states (and some states prohibit people from owning a still for any reason). The key takeaway is this: it's illegal in the United States to do this ourselves at home. Maybe in the future apple brandy will join beer, cider, and wine as something permitted in moderation for home consumption across the United States.

It is not illegal, however, to understand how these delicious things are made, nor is it illegal to go out and buy them, thankfully (well, at least in most places in the United States). In fact, if you go out and buy a commercial young apple brandy and combine it with fresh apple juice, you have the makings of your own pommeau. Because of its official designation, you can't really call it pommeau unless you are in fact living in Normandy or Brittany in France, but hey, it's probably even less likely that a Frenchman will show up at your door to

complain than someone from the TTB will show up looking for your still, so feel free to name your homemade beverage a pommeau if you want.

The hardest part of our two-ingredient pommeau recipe is not combining two liquids and mixing, but rather waiting the many months and, maybe, many years for it to age to perfection. Here is a little trick that we have found helpful: before pressing your apples for the year, gather your supplies. In an ideal world you would have a used 5-liter oak barrel for aging your pommeau and a couple of bottles of apple eau de vie, apple brandy, or applejack. However, there are some considerations before you run out and buy a new barrel. One is that small barrels expose more of the liquid within the barrel to the wood because, by volume, more of it is coming in contact with the surface area in the barrel. This can cause two issues. The first is an overly woody flavor that dominates the delicate apple flavors of the pommeau; a well-used barrel solves this problem, and if you do buy a barrel, it will get better every year. The second is that in a dry climate that volume can evaporate out; this is harder to overcome. We have included a no-barrel recipe just for this reason.

After you decide whether or not you can use a barrel, the pommeau begins at pressing time. Reserve a gallon of juice and make your own pommeau using the recipe that follows. When you are done, tuck that little barrel away somewhere cool where you won't see it too often. With luck, you will discover it the following year when you make your next batch of cider and you will make a mental note to check back on it in the New Year. Think of it as a really big cocktail that you mix up but can't drink for 15 months or longer.

Pommeau

By adding sufficient alcohol to fresh or lightly fermented juice, we prevent the fermentation process from starting, killing the yeasts and thereby leaving all those sugars for us. The alcohol also serves as a preservative, which is handy since we want to tuck this away in a wooden barrel or glass jug to age for at least 15 months. The final ABV percentage is the result of diluting the brandy's alcohol level, plus any alcohol that was created if you use juice that has begun to ferment. Once bottled, this just gets better with age, so we suggest bottling in smaller (350 mL) bottles and tucking them away in odd places to be discovered in the years to come.

If your barrel is brand-new, you don't need to clean it; jump to step 2. Your first-year pommeau will be a bit oaky (or whatever wood you are using), but after that it will get smoother and smoother. If your barrel contained your last batch of pommeau that you just bottled, you can skip steps 1 and 2 and go directly to filling it. If your barrel contained something else, like wine or a different alcohol, you probably want to clean out any residue so that your pommeau has every opportunity to shine. And if you can't find a barrel, fear not — you can still make pommeau (see page 277).

**YIELD: ABOUT
1 GALLON**

1 gallon (3.8 L) fresh
or slightly fermented
sweet cider

2 (750 mL) bottles apple
brandy

1. If necessary, clean the 5-liter oak barrel by filling it about one-third full with hot, chlorine-free water, and plug the bung securely in place. Give your barrel a shake for 30 seconds or so. Open the bung plug and empty the water from the barrel, paying attention to the color of the water. Give the barrel a sniff. If the water was discolored by the previous liquid or you smell a lingering scent of it, go ahead and repeat the process, this time letting it sit for an hour in the barrel before giving it another shake and pouring it out.

2. Rehydrate the barrel by filling it about one-quarter full of hot, chlorine-free water. Give the barrel a good shake for 30 seconds or so. Set the barrel on one of its heads, fill the head with hot tap water, and let it sit for 30 minutes. Carefully pour the water out and turn it over to set it on the other head. Fill the head with hot tap water and let it sit for 30 minutes. Pour the head water out and set the barrel on its side with the bung facing up. Fill the barrel to the top with cold chlorine-free water and look for leaks between the stave joints. If you think you see some, top off the water and leave it in the sink for 24 hours. Pour the water out of the barrel. The barrel is now ready for your pommeau.

continued on next page

Pommeau
continued

3. Mix together the sweet cider and brandy in a clean 2-gallon or larger container, stirring slowly. Place a clean funnel in the bung hole of the barrel and pour the cider-brandy mixture into the barrel until it reaches the top. Any remaining liquid is in the domain of the cidermaker (this is not an excuse for increasing the recipe ingredients or finding a smaller barrel). Plug the bung securely in place.

4. Set the barrel in a cool place out of direct sunlight and age for 15 to 18 months.

5. Carefully open the bung and use a wine thief to draw out a sample. If the color and flavor is agreeable, it's time to bottle. If it seems a bit harsh still, you might want to close it up and age it for another 3 months.

6. Using a clean raking cane, siphon the pommeau into clean bottles and store for as long as you like.

STREUOBSTWIESEN:
BIODIVERSE GERMAN APPLE ORCHARDS

In many parts of the world, apples are farmed like row crops, with ever-shrinking trees planted in tightly packed rows, perfect for high-tech picking equipment. The ground in between, also known as soil, is kept clear so that the giant apple shakers and sweepers can crawl their way methodically down the rows.

There are, however, holdouts that remind us of the idealized orchards of our dreams, where either annual crops are grown in the alleys between apple tree rows (known as alley cropping) or pasture is managed and livestock are carefully moved through for grazing to control the grass (known as silvopasture). In Germany they are called Streuobstwiesen, and they are designed for multiple uses and biodiversity. The trees are on large stock and thus can live for more than a century. The grass meadows between the trees are grazed by cattle and sheep, while the ecosystem supports a wide range of native plants and animals. As pressures mount to more efficiently utilize the land, conservation groups are stepping in to help consumers understand and support the orchardists and cidermakers continuing this tradition.[20]

No-Barrel Pommeau

If you can't find, or would prefer not to use, a small barrel in which to age your pommeau, you can use a 1-gallon glass jug and add a wood aging spiral or inner stave. This will give you the flexibility of controlling the amount of wood (barrel) flavors to your taste. Like the barrel, these are reusable and will also mellow over the course of subsequent pommeau batches.

YIELD: 1 GALLON

---

3 quarts (2.8 L) fresh or slightly fermented sweet cider

1 (750 mL) bottle apple brandy

1 wood, oak, or other choice spiral or inner stave (see Suppliers, page 321)

---

1. Pour the cider and brandy into a clean 1-gallon glass jug. Add the spiral or inner stave by tying a string to the wood and securing the other end of it to the neck of the glass jug. Fasten and tighten the cap, and slowly swirl the jug to incorporate the cider and brandy.

2. Set the jug in a cool place out of direct sunlight and age for 15 to 18 months.

3. After aging, pour a small sample and taste. If the color and flavor are agreeable, it's time to bottle. If it seems a bit harsh still, you might want to close it up and age it for another 3 months.

4. Using a clean funnel, pour the pommeau into clean bottles and store for as long as you like.

POMMEAU   NO-BARREL POMMEAU

Bourbon Cider

This recipe came about by wondering just how you could reach the alcohol limits of typical commercial yeast. We learned that dumping in all the sugar at once often didn't work — it probably overwhelmed the yeast with too much sugar and not enough nutrients. When we spaced out the sugar it worked, but it resulted in a cider that wasn't that great flavor-wise. By adding the bourbon-infused oak chips and then tucking it away for months, we found something that consistently surprises and delights.

YIELD: 3 GALLONS

½ cup (118 mL) unchlorinated water

1 teaspoon champagne-style yeast (Lalvin EC-1118, Red Star Premier Curvée or Premier Blanc)

9 cups (1.8 kg) granulated cane sugar

3 gallons (11.4 L) preservative-free apple juice

4 ounces (112 g) oak chips

1 cup (237 mL) bourbon

1 (12-ounce/350 mL) bottle dry cider

1. Sanitize a 3-gallon carboy, 5-gallon bucket, funnel, stainless steel whisk, and an airlock with a no-rinse sanitizer.

2. Heat the water to 104°F/40°C and pour into a quart canning jar. Sprinkle the yeast over the hot water, stir gently, and let it sit for 20 minutes. Stir again, measure the temperature of the yeast mixture, and write that down in your cider log or on a piece of masking tape attached to the 5-gallon bucket.

3. Mix together 6 cups of the sugar and the apple juice in the 5-gallon bucket and stir with the whisk until the sugar is dissolved. Measure the SG of the juice-sugar mixture with a hydrometer and the temperature with a thermometer and record them in your cider log or on a piece of masking tape attached to the bucket.

4. If the temperature of the yeast and the juice are within 18°F/10°C of each other, proceed to step 5. If not, add ½ cup of the juice to the yeast mixture, gently stir, and wait for 5 minutes. Measure the temperature and if it's within the 18°F/10°C range, move to the next step. If not, add 1 cup more of the apple juice to the yeast mixture, stir, and wait for 5 minutes.

5. Pour the yeast mixture into the 3-gallon carboy. Add some juice to the yeast jar, swirl it around to get all of the yeast remaining in the jar, and pour it into the carboy. Add enough of the juice-sugar mixture to the carboy to fill within 3 or 4 inches of the top. Place a piece of plastic wrap loosely over the opening. Reserve any remaining sugared juice and refrigerate for later use.

6. Place the carboy on a surface that is easy to clean or set it on a tray or pan. Let it sit in an environment where the temperature is between 55°F/13°C and 65°F/18°C. Bubbles will slowly form after a few days, then build. If the liquid froths out the top and down the sides, simply clean the sides and tray/floor with a wet sponge.

7. When the bubbles stay below the top of carboy (this can take 3 to 10 days or sometimes more), apply the bung or lid, insert the airlock, and fill to the appropriate level with either fresh water or a neutral distilled spirit.

8. The primary fermentation is finished when half to three-quarters of the sugars have been consumed, which should take a couple of weeks given how high the sugars are in the beginning of the primary fermentation. Taste the cider used to measure the SG and write any tasting notes in your cider log or on a piece of tape attached to the carboy. Use a wine thief to draw out about 2 cups of cider and place in a clean quart jar. Stir in the remaining 3 cups of sugar until dissolved. Pour this mixture back into the carboy and replace the airlock, refreshing its water if necessary.

9. Sanitize another 3-gallon carboy, a racking cane, and a siphoning hose.

10. Using a siphon, rack your cider off into the carboy, making sure to draw off all the cider above the lees, without drawing the lees out.

11. Place the oak chips in a quart jar and add the bourbon, then stir to incorporate. Secure the lid. Let sit for 1 week, giving the jar a gentle shake every day to make sure all the chips come in contact with the bourbon.

12. Take another SG reading to determine if the cider has returned to the desired range of about 1.030 or lower. If it has, then proceeded to step 13. Otherwise note the SG on the label or in your cider log and keep checking until it reaches the desired range.

13. Strain out the wood chips from the remaining bourbon and drop into another sanitized 3-gallon carboy. Using a siphon, rack your cider off into the same carboy, making sure to draw off all the cider above the lees, without drawing the lees out.

14. Add enough bottled cider to top off the racked cider to within a couple of inches from the top of the new carboy to minimize air contact. Reapply the airlock and fill it to the appropriate level with either fresh water or a neutral distilled spirit.

15. Leave the cider on the oak chips for at least 1 week, or however long it takes to create the flavor you desire, tasting every day or two. A warning: It can turn from almost there to too much very quickly, so don't wait too long between tastings. We have found that about 10 days is just about right for our tastes.

16. Siphon the cider into clean bottles, secure the tops, and store for at least 6 months to allow the harshness of the higher alcohol levels and the oak to mellow. The bottles can be stored in a cool environment out of direct sunlight for 2 years or more, given the higher alcohol levels.

JACKING CIDER:
AN AMERICAN TRADITION

Cider was once the king of alcoholic beverages in colonial America. And even then, there were those who wanted their cider a little harder. It was said that wherever there was an apple orchard and a clear, cold stream, there would surely be a still running somewhere nearby.

For those in the northern, colder climates, there was an easier way to jack one's cider than running a still. Many simply left their barrels of cider in the barn to repeatedly freeze and thaw over the winter, by which time the sweet cider would have likely finished fermenting. This is known as fractional crystallization, which works because water and ethanol have different freezing point temperatures. Each time the cider would begin to freeze, the water would freeze first. As the ice was removed, there was an increasingly stronger concentration of ethanol in the remaining cider.

While simpler than building and running a still, there are two drawbacks to this technique. First, the alcohol percentage doesn't come near to what you can get from distillation. At best, you can (theoretically) get something around 30 percent ABV, while you can get more than double that with distillation. Second, unlike distillation, where the impurities are drawn off from the beginning and ending of the distillation run, with the fractional crystallization process those impurities are concentrated in the liquid that remains. This unfortunate trait was widely regarded as responsible for the condition referred to as "apple palsy," and it sounds like it was pretty bad. The *New York Times* put it this way in its April 10, 1894, issue: "The victim of applejack is capable of blowing up a whole town with dynamite and of reciting original poetry to every surviving inhabitant."[21]

At this point it would be easy to think of applejack as the rustic country cousin of a more refined apple brandy, or Calvados if made in the proper regions of France, and that's basically what we thought, but there were a couple

of things that didn't fit. First, on the brandy shelf of our local liquor stores, beside amazing local apple brandies and official Calvados from France, there was a humble bottle of Laird's Applejack with a plastic screw top and a price tag that was half the size of the others. It turns out that Laird is the oldest distillery in the United States, and they don't rely upon the weather to produce their products. Officially, the TTB defines applejack as a grain-neutral distilled spirit that is flavored with at least 30 percent apple brandy. Freezing barrels are romantic and appeal to a simpler way of life, but they have nothing to do with the applejack available to you today. The simpler version is, however, much more accessible to the home cidermaker.

BEING STILL

Stills have an interesting place in our family history. Christopher's Midwest relatives of a few generations back were said to have had a knack for producing fine moonshine that was undoubtedly helpful in getting them through bone-chilling winters and hot, humid summers. By Christopher's generation, the skill was no longer being passed down in the family.

For an all-too-brief period of time, our family farm had four generations living on it, the eldest being Kirsten's mother, Nadine, and stepfather, Hillman. Nadine was a pure lifelong learner who could go from hearing about something, to an Amazon order, to up to her elbows in the process within days. That's how we came to be the only farm in the area (we think) to have not one but two pot stills in working order. She was more interested in making botanical essential oils than making a fine brandy, and though she and Christopher planned to combine their passions and skills to try a farmstead apple brandy, it never happened before Nadine passed.

Both of Nadine's stills are beautiful, hand-pounded copper pot stills, which are known as alembic-style or French-style. The other style is a column distiller or German-style. One of the big differences is that with column distillers, you can achieve higher-proof spirits with one pass, while it takes a couple of passes with the pot styles. The column style also tends to better preserve the fruit essences so, with the exception of apple brandy, it is typically the still type used for fruit brandies.

While it is true that the Internet and some cider books offer everything from advice to full do-it-yourself plans for building a kitchen still, we won't here because besides being illegal, Christopher would be the one to build it, and honestly, he is just not that handy or precise to build something that, if done incorrectly, will explode. A general rule on the farm of staying away from big things that could explode has served us pretty well over the decades, as we are both still here and nothing has exploded except for a big bottle of ginger bug (naturally fermented ginger ale) and a few highly carbonated ciders in the cider cave. There was also the time our teenage boys lit a giant bonfire in a far field with flaming arrows and gasoline, but that was completely unsanctioned.

Applejack

This recipe differs from the Ice Cider on page 261 because with ice cider we freeze and thaw the fresh juice to separate the sugars in a concentrated form and then ferment. In this recipe we go the other way by first making the normal cider and then freezing and thawing to separate the alcohol in a concentrated form. How concentrated? This process will not get all of the alcohol out of your cider, so in effect, you will be leaving some of it locked up in the ice that hasn't melted.

Here is a crude guess as to the alcohol content in your finished applejack: A gallon of cider at 7 percent ABV has 9 ounces of alcohol in those 128 ounces. Assuming you got all the alcohol in the half gallon you pour off, you would have something in the range of 14 percent ABV, or double what you started with. The less water you also siphon off, the higher your ABV, but know it's very hard to get all of the alcohol and it will always include some water. Hence, while fun, it's time consuming and not a path to brandy.

YIELD: ABOUT 2 QUARTS

---

3 gallons (11.4 L) cider (5–9% ABV)

---

1. Pour the fermented cider into a clean 5-gallon food-grade plastic bucket. Place in a freezer with the lid either loosely fitting or without the lid (chest freezers work best, but you can use a stand-up style with the shelves removed). Let freeze for 18 hours.

2. Check on your cider. If a layer of ice has formed on the top of the cider, pull the bucket out of the freezer and carefully break and remove the ice. Return the cider to the freezer. If ice hasn't yet formed, check again in 6 to 12 hours and remove the ice.

3. Repeat this process for 5 to 7 days, until ice is no longer forming on the surface. At this point it is ready to be carefully siphoned off and bottled.

CHAPTER 9
DRINKING CIDER

Is there a right way to drink cider? Yes, of course there is. It's just that the way you like to drink it might be different from the way we do, and neither of us is wrong nor right to our way of thinking. This point is illustrated by a story Bill Bleasdale (page 143) told us about a cider competition he attended. As the experts gave their analysis of his cider, listing both its virtues and its faults, Bill said it occurred to him that some of the "faults" he considered to be virtues, and some of the "virtues" were things he didn't like. Bill's story reminded us that taste is indeed subjective.

Christopher trained as a cider judge and started volunteering in competitions in 2019, so in this chapter we give some simple cider tasting hints that he picked up, as well as some insights into the science of flavor. In the end, though, it's about what you like. If it brings you joy, it must be a good thing in our book. Our biggest recommendation is to keep stepping outside what you know and try those styles that you don't. Enjoy the cider renaissance!

those in context — and where apples come from — was just brilliant." As the cidermakers explored multiple parks where the trees are protected, they were surprised to discover both hops and cannabis growing as companion plants. The park officials explained that the hops had been introduced, but it appeared the cannabis has been a companion plant for a long time.

We had planned to spend a good part of our conversation discussing Peter's demonstration cider operations, which he has run for a decade or more, but we quickly learned that he had recently sold it. He plans to carry on the tradition of research performed by the famous Long Ashton Research Station and conduct much-needed cider research. "Most people at my age buy a fast car, but I bought a laboratory," Peter explained.

Our conversation ended by discussing the rise in popularity of geography-based cider styles, like Spanish and French, and the versions of these popping up commercially here in the States. "These things are best enjoyed in context. The best place for Spanish cider is in Spain . . . The best place for French cider is in Normandy and Brittany," Peter said. Cider that has developed in a specific region

has also developed in and around that region's culture. Peter said the cider cultures of the Asturias and Basque Country are exceptional and quite fun to experience, so his advice is — go there, immerse yourself in the culture, and enjoy these ciders as part of that culture.

THE SCIENCE OF FLAVOR

Now that you've received some hints for how to be a better cider taster, you may be wondering: *Where does the flavor I am tasting come from?* Flavor is technically the combination of three characteristics: taste, aroma, and texture.[22] While cider's nonvolatile compounds (things that tend not to vaporize) are registering on our tongue and in our mouth, its volatile compounds (things that tend to vaporize) are registering in our olfactory regions, while the tannic combination of astringency and bitterness is providing the feeling of texture. All this is sent to our brain, which stitches it all together to put words and emotions to what we are tasting.

We are going to look at four elements that cross volatile and nonvolatile compounds: acids, aromas, sugars, and phenolics. Just remember AASP. (You could think of it as the new form of AARP, American Association of Sexy Persons.) Onward. With a little knowledge of chemistry and microbiology, you are going to understand the underlying structure that presents those crazy butterscotch, black tea, forest floor, or pear skin flavors we see on the labels and come to recognize ourselves.

Acids

Sour flavors come from acids. Humans have perception thresholds, whereby we don't notice something until it reaches a certain point, which usually starts with pleasure. Then the levels rise to another point where we perceive something as too this or too that and it ends with pain. In the case of acids, at the minimum threshold we can begin to taste the sour and at the upper end we begin to be repulsed (well, most of us) by the high level of sour.

Different acids also taste differently to us. In wine, the main acid is tartaric acid, which tastes a bit like salty Meyer lemon. Tartaric acid is absent in cider. The main acid in cider is malic acid, which tastes somewhere between a Granny Smith and an unripe apple, and for some people it has a bitter metallic finish. Acid tastes can also be hidden from us, to varying degrees, by the levels of sugar and alcohol that are present. As you might be guessing, we have perception thresholds for sugar and alcohol as well.

The amount of malic acid in the beginning of the fermentation process is directly attributable to the apple varietals that are used. As the series of microbes come into play, the amount of malic acid can be increased by the actions of the yeasts (fixation of carbon dioxide by pyruvate to make oxaloacetate, which then reduces to malic acid) and decreased in the malolactic conversion (mostly driven by lactic acid bacteria) or by *Saccharomyces cerevisiae*. When you drink cider, the level of sourness you taste is mainly due to its acids.

Aromas

There are, perhaps, few things more lovely in a pub than the fruity and floral aromas waiting for you in a glass of really good cider or perry. (Note we do qualify that as "in the pub" and recognize there are plenty of lovely things in the world that would rank above the aromas of a good cider, but not so much in a pub.)

Scientists estimate that 75 to 95 percent of what we taste is actually what we are smelling.[23] You will smell that lovely imagined pub cider or perry not once, but twice. The first time is through your nasal passageway when you lower your schnoz into the glass and breathe in. This is officially called "orthonasal olfaction," in case you want to impress someone at your next tasting. The second time you smell this cider or perry is when you "chew" it, which you might think of as drinking, but stay with us. When you take anything into your mouth, your teeth, tongue, and cheeks are wired to break it down into a size easily assimilated later in its journey through you. In the case of a drink, the teeth can give it a rest, but the aromas still follow the second smelling pathway, that of the retronasal olfaction. Basically, as you

swallow, the cider's aroma molecules slide down the roof of the mouth. At the throat, those aroma molecules part with the cider and head north to your olfactory glands instead of south to be digested. This is why slurping cider helps you to "taste" it — you are increasing the flow of aromas to your retronasal olfaction processing. Remember that phrase when someone gives you a dirty look after a particularly loud slurp.

Sugars

Sweet flavors come from sugars. Microbes love sugar. In fact, you could view fermentation as an unperceivable number of microbes chasing sugar. If any of the apple's original fructose (74% and the sweetest), sucrose (15% and in the middle of the sweet scale), or glucose (11% and the least sweet)[24] remain at the end of fermentation, it wasn't by the microbes' choice. The microbes likely met their fate prematurely in one of two ways: they died at the table through a human intervention, like sulfur dioxide applications; or they succumbed to the defense mechanisms of other microbes, like increasing the acidity in the room or producing biotoxins. If they didn't die from either of these ways, then they died from the alcohol, which, as we all know, turns the lights out permanently in large enough doses. If the microbes are left to it and the cider is finished dry, with no fermentable sugars remaining, the cidermaker must add sugar back in at bottling, when no microbes remain, if he or she wants to produce a sweet cider.

If you are making perry, be aware that there are some types of sugar alcohol, like sorbitol,

Cider's color has a lot to do with how apples are ground and pressed, and specifically how long the pomace and must are in contact with air. This process is also known as enzymatic browning, which gives us a hint as to the color element. It's the enzymatic oxidation of the phenolics that gives us the varying shades of cider — from light straw to deep butterscotch. The enzymatic oxidation of the apple's phenolics creates new phenolic molecules as well, many of which have no color to contribute to the cider but they do contribute antioxidant traits.[25]

that are "unfermentable," meaning that microbes cannot process them, and they are therefore left for us (there is only a trace amount of sugar alcohols in cider). In fact, it takes a longer time for the microbes in our gastrointestinal tract to process sorbitol as well, and at amounts of 50 mg or more per day it may have a laxative effect. While sorbitol is half as sweet as sugar, it may increase the viscosity in the cider. We have heard stories of imbibing too much perry leading to an unintended sorbitol-induced GI reaction.

Phenolics

Plants contain chemicals — literally thousands of chemicals across the plant kingdom. You might have heard the term *phytochemical* when reading about antioxidants or phytoestrogens; this word simply means "chemicals found in plants." Phenolics are a class of these phytochemicals that share chemical makeup and some health benefits for humans. Two types of phenolics that have a direct impact on the flavor of ciders are the "naughty twins" (as we refer to them): astringency and bitterness. Astringency provides the texture of the flavor we perceive, and bitterness is one of the five basic tastes. To say a cider apple variety is "tannic" really means that it has higher levels of astringency and bitterness, which are similar but different.

We experience astringency in the entirety of our mouth, including across our tongue, as a puckering and drying out of the mouth. It feels as if all the wetness in our mouth has been magically and instantly drawn out — leaving us with a dry, furry, uncomfortable feeling. Think unripe fruit, especially persimmons. Procyanidins are one of the groups of phenolics in cider and are polymers of catechins, which are antioxidants. Procyanidins with a high molecular weight (longer chain molecules) are responsible for astringency. Their magic saliva disappearing act has a scientific explanation behind it: hydrogen bonding takes place in your mouth between o-diphenolic groups and salivary proteins, causing them to turn to a solid. That's what you feel is coating your mouth.

The other naughty twin is bitterness. Unlike astringency's over-the-top performance in our mouth, bitterness is more of an intense taste on the tongue. Procyanidins with lower molecular weight (short-chain molecules) are responsible for bitterness. Our reaction to bitterness is one of our most basic senses — and for good reason. While many people don't like to eat plants that much, the feeling goes both ways because plants don't like to be eaten either. They like to reproduce, just like us. They protect themselves as best they can with a mix of toxins that often have a very bitter taste. If you come across a new plant on your walk and put it in your mouth, it's likely that an intense bitter taste will make you involuntarily spit it out, perhaps saving you from a poisoning for that day. Our same basic wiring is in place when you taste a perry pear in the field or some very tannic cider apples. It's acceptable to spit in the orchard, but usually not in the pub.

Yeast as Flavor Makers

If you have ever leaned into a freshly opened carboy, imagining tropical fruits or warm spices, only to be assaulted by aromas from a nail salon, you have experienced the overproduction of one of the organic compounds produced by yeast in the fermentation process: ethyl acetate. This ester can make up the majority of esters in cider. Esters are what give us the pleasant things, like fruity and floral aromas, but when there are too many of them, you go from tropical fruit to paint solvent — a pretty wicked transition. Researchers have pinned the threshold where things turn from nice to not so nice at 200 mg/liter.

Ethyl acetate is produced at the beginning of the alcoholic fermentation and remains stable through that phase.[26] Different yeasts produce differing levels of ethyl acetate, which is one of the reasons S. cerevisiae species are popular among cultured yeast strains — they produce low levels of ethyl acetate. If your wild-fermented cider ends with nail polish remover tones (also known as excessive EA), you need to bring down the amount of ethyl acetate to below 200 mg/liter. In fact, humans can only perceive ethyl acetate at about 120 mg/liter. By blending the cider with another low-EA cider, you can effectively bring down the amount to either below the drinker's perception or at least to a level that is perceived as pleasant.

Another organic compound produced by yeast that affects flavor is higher alcohols — compounds like isobutanol, propanol, and butanol. These are not the same as the alcohol (ethanol) that we are trying to produce. An easy way to remember the difference is that higher alcohols have a higher number of carbon atoms; ethanol has two, so all the higher alcohols have three or more. Like with ethyl acetate, humans have a perception threshold for higher alcohols. Research shows that at concentrations beginning at 300 mg/liter, we perceive a pleasant flavor, but at higher concentrations — around 400 mg/liter —that turns to unpleasantness. Some compounds have harsher taste profiles than others.[27]

Carbonation as a Flavor Conveyer

The gas responsible for the effervescence in cider is carbon dioxide (CO_2), which can occur naturally in cider or it can be put there by us. Both methods produce effervescence, but there are several differences in the resulting quality of carbonation that you should be aware of when deciding how you want to carbonate your creations.

Carbon dioxide is naturally produced by the yeast during both primary and secondary fermentation, but it's the secondary fermentation in the sealed bottle that produces the pop. According to Henry's law, an equilibrium is reached between the carbon dioxide dissolved in the cider and the carbon dioxide floating about in the headspace between the cider and the cork, bail top, or cap. Ideally, the amount of carbon dioxide produced by the yeast in your closed bottle is just enough to produce a pleasing pop upon opening. Too little sugar or yeasts at bottling, and you get very little to nothing upon opening — like you bottled tap water. Too much sugar and yeast, and the party in the bottle gets out of hand. The yeast produces ethanol and carbon dioxide, building great pressure above and below the cider line in the bottle, to the point where opening your cider becomes an event best attended in rain gear. (Think of those Indy 500 race-car drivers with their shaken giant bottles of champagne spewing forth over everyone.)

A cidermaker may inject carbon dioxide into the still cider, a process referred to as "carbing," to better control the level of effervescence in the finished cider and avoid the extremes of the above natural process.

When the seal on the bottle is released, the carbon dioxide in the air gap is immediately released and the balance in pressure becomes unbalanced. The cider becomes "supersaturated," meaning there are more carbon dioxide molecules dissolved in the cider than could be under normal atmospheric conditions, which it is at now with the top popped. Nature wants its balance back, and to get it the excess carbon dioxide needs to be released out through the bottle opening and into the atmosphere. For the carbon dioxide to diffuse at the surface of the cider, it needs to get up to the surface in the form of bubbles.

Bubbles seem magical, but they don't just magically appear in your cider; they form through a process called nucleation, which is pretty cool. The carbon dioxide needs a cavity of some sort to form into a microbubble, which is less than a micron in size — about the size of a bacterium. Glassmakers know this and use lasers to etch minute dimples in the bottom of some types of glasses. If you see a perfect column of bubbles

Yeast produces ethanol and carbon dioxide, building great pressure that is released when the bottle is opened.

in your cider poured in a champagne flute, it's likely intentional. There is a much more common and predictable place for baby bubbles to hatch than laser etchings — lint. The next time you see a bored bartender drying glasses with her bar towel, think of the minute fibers (the fibers that are released are roughly 100 microns long) being planted in that glass. When your cider is poured into the glass, the towel fibers develop tiny gas pockets. These pockets are easier places for the carbon dioxide to build a bubble, so as soon as the bubble reaches 10 to 50 microns it releases from its lint home and a new bubble immediately begins to form in its place.

These bubble wombs in your glass can be very productive. In physicist Gérard Liger-Belair's book *Uncorked: The Science of Champagne*, he reported finding as many as 30 bubbles per second being produced from one site. Each one is forming, attracting other carbon dioxide molecules, and growing until it reaches sufficient size to detach from the surface. Once free, the carbon dioxide bubbles must find each other and continue to group together in order to resist the forces of the cider keeping them down. As the group grows, it begins to win the battle with the surrounding liquid molecules' relatively weak electrical bonds and it rises upward toward the surface of the cider. When we look at a bubbling glass of cider, we are seeing these carbon dioxide bubble clusters pushing their way through the cider molecules toward the exit.

Christopher likes to imagine a stadium full of people at a concert, all wearing the same golden cider-colored T-shirts. As the lights come up and the stadium is filled with pulsating music (cork is popped), hundreds of new fans appear sporadically around the periphery of the venue in their light blue carbon dioxide T-shirts. They immediately push through the cider fans to find one another and lock arms. As each group grows, they begin to push through the crowd, breaking whatever bond they had with their neighbor or with the band. Other carbon dioxide mini-crowds follow, and soon there are hundreds, then thousands of them streaming for the exits. When they finally make it to the surface, they pile up — think of the foamy head on a beer, or the typically smaller and thinner head on a carbonated cider. This isn't the end of the carbonation story, and from a taste perspective it's actually closer to the beginning. That's because up until this point carbonation has been all about the visual show for us. Bubbles are beautiful and mesmerizing, especially if you have had a few pints of cider already. Now that they are piled up at the surface of your cider, the flavor work begins.

You might be asking: What does this have to do with flavor? Carbon dioxide affects your sense of flavor in two ways. First, when the bubbles pop on the surface of the cider as you take a sip, they are actually engaging your olfactory receptors. Remember the orthonasal olfaction you just learned about in the aromas section above? To understand this, we need to understand how bubbles burst. When a bubble's film bursts, a cavity is formed where the carbon dioxide was just before. Cider likes to keep her surface clean and neat, so minute cider waves rush in from the perimeter of the cavity and collide in the middle, pushing up a tiny geyser of cider. This stream then breaks apart into tiny droplets that remain suspended in the air, ready for us to breathe in. You would be correct in saying that it is your nose, before your mouth, that takes the first sip.

The second way carbon dioxide affects your sense of taste is with its sting. When something stings, we experience something called chemesthesis. The nerve that carries the message of pain in our mouth to the brain, the trigeminal nerve, is engaged, which heightens our senses.[28] Perhaps that's an argument for either drinking your cider still or carbonated, depending upon how many heightened senses you are looking for.

WANT TO GO FROM HOBBY TO MICRO-CIDERY?

The cider industry and community in the United States is young and defining itself. It isn't held to or bound by traditions, which can't be said of the older U.S. beer and wine industries or cider industries abroad. It is exciting to be a part of how it grows and expresses itself, in many ways because there is a chance to build a community of makers that is contemporary, inclusive, ecological, and egalitarian. If you want to grow your cider hobby into a business, check out our downloadable guide at https://www.storey.com/cider-business/.

cider, beer, or wine yeast nutrients are a mix of a number of nutrients, but if you are using straight DAP, we would recommend adding the amount recommended on the label. Most suggestions will fall around ⅓ to ½ teaspoon per 3 gallons of cider.

SLOW OR STUCK FERMENTATION (WILD YEAST)

Symptoms. The symptoms for a stuck wild yeast fermentation are the same as the commercial yeast variation — nothing is moving.

Cause. In the case of wild yeasts, it could be that they are just taking longer than their commercial cousins, which, remember, have been carefully bred for characteristics like active fermentation levels. Wild yeasts are often slow growing, so that means weeks for a primary fermentation and months for a secondary fermentation.

Resolution. If you don't have the time or patience for the longer fermentation of wild yeast, then try a batch or two of commercial yeast and compare the results. If you can't tell the difference or prefer the results you have gotten with the commercial yeast, then we suggest sticking with what works for you and your cider adventure. If, however, you like the taste the wild yeasts produced, then perhaps reset your expectations for how long it will be until you are enjoying your cider creations. When using wild yeasts it's not uncommon to wait 6 to 9 months from pressing until you are opening your first bottle. A tip is to stagger production throughout the year, so you always have a vintage ready to enjoy.

SULFUR IN MY CIDER!

Symptoms. There is an unpleasant aroma that varies between rotten eggs, burnt rubber, and that smell when your kitchen sink starts to back up. Although it's natural to catch a hint of sulfur during fermentation, this is something more — it is quite memorable and very unpleasant since it's originating from your cider.

Cause. Hydrogen sulfide (H_2S) is squarely to blame for this one. There are at least three possible causes. It

could be the sulfur caught a ride on the apples, from an application or two of sulfur in the orchard to fight fungus. Even if the apples are grown organically, copper and sulfur are allowed as a fungicide. It might also be that your yeasts were stressed out. This can happen when they haven't been able to get enough nutrients and sugars for proper growth. Yeast can also produce excess hydrogen sulfide if the fermentation was at too high a temperature.

Resolution. This too may pass, or not. This has only happened to us once and we noticed it at bottling time. Not knowing what it was, we bottled quickly to keep the evil genie in its bottle and marked them with a big double XX on the label. We opened one in a few weeks and it seemed to have angered the sulfur gods within, so that one got dumped down the drain. About 9 months later, when the cider cave was looking pretty sparse, and after having one or two ciders, our courage and curiosity were boosted enough to open another XX-branded bottle. We thought it was mislabeled — it was fine. Not great, a little thin, but no rotten eggs poured forth. The other two were the same.

Other resolutions depend upon where you are in the fermentation process. If still in the primary fermentation stage and you have yeast activity, then feed them! You can find nutrients at beer and winemaking supply stores; just follow the directions for the amount of cider you are working with. If you are racking between primary and secondary fermentation, you are going to break the rule of minimizing the amount of oxygen you introduce to your cider when racking. Instead of keeping your siphon tube at the bottom as it fills, below the cider line, you are going to do the opposite. With your primary cider carboy elevated clearly above your secondary carboy, siphon while keeping the end of the tube just inside the carboy neck, allowing the cider to drop from there to the cider line as it fills. You will see a froth and you will smell the smell, but that's good because what you are doing is airing it off. Add the airlock and give the secondary fermentation a few weeks before giving it a sniff. If you detect faint traces or nothing, then you should be homefree. If it's still there, then it's time to find a penny and a friend.

Dig into your change jar and find the cleanest penny you have that was made in 1982 or earlier. Those pennies are made of pure copper, and that's what we are looking for because copper combines with sulfide molecules to form copper sulfate ($CuSO_4$), which is insoluble, so it settles to the bottom with the lees. Pour the affected cider into a widemouthed pint jar and drop the clean penny in. Give the cider a slow stir with a spoon for a few minutes, then blow across the top of the jar to remove any aromas lingering on the surface of the cider. Remove the penny, then pass the cider to an unsuspecting friend to smell and taste. Ask them what he smells and what he tastes, and, most importantly, don't tell him beforehand why you are doing this. Why? Because our brains are great at telling our senses what they sense, even if it's not what they sense. Your brain is on high alert because you don't want a carboy of rotten egg cider so it's going to be keen to sense if it's still there, but your unsuspecting friend's brain is not on high alert. Hopefully he's trying to remember all the tasting terms you once taught him and rotten eggs is probably not one of them. If that smell is still there, you will see the puzzled look in his face, perhaps an involuntary wrinkled nose and slight recoil away from the jar. If so, the hydrogen sulfide is bound too tightly for the following copper treatment to work and your best bet is to bottle it and hope it dissipates over time. If, however, you get a list of pleasing tasting notes that do not include rotten eggs or septic system, then read on for a resolution.

A well-known way to correct this problem in wine is to add copper sulfate (though some today are questioning its effectivity), and that's what we can do with cider as well. You need a tiny amount of copper, 1 part per million. This equates roughly (you need to do the numbers exactly for your amount of cider) to 1 teaspoon of copper sulfate diluted into 2 quarts of water. One teaspoon of that dilution goes into 3 gallons of cider. As with all measurements, it's a lot more accurate to stay metric through this math but that's the price we pay in America for stubbornly holding on to our unique measurement system, isn't it? There are also commercial products like Kupzit, which is copper citrate coated in high-quality bentonite clay pellets. You can source these in fairly

small amounts, but again, the numbers can be difficult to get down to the proper measurement for your home-sized cider (like less than $\frac{1}{10}$ of a teaspoon for your 3 gallons of cider). If you are planning a micro-cidery, then you will be dealing with larger amounts of cider at risk, and so we would recommend going with the prepared copper citrate route based upon our conversations with commercial cidermakers.

CIDER SICKNESS (A.K.A. LA TOURNEY, A.KA. FRAMBOISE)

Symptoms. Your cider is hazy to milky opaque with an aroma that has been described as lemon or banana skins. In France, this disorder is called framboise, which means raspberry, because it can also take on a raspberry odor.

Cause. Cider sickness is caused by the bacteria *Zymomonas mobilis*, although in the mid-1950s scientists identified another strain, *Z. anaerobia* var. *pomaceae*, as the troublemaker.[29] These bacteria feed upon fermentable sugars and produce acetaldehyde, also known as ethanal. Conventional wisdom, including that given in nearly all cidermaking books, is that *Z. mobilis* can only take hold in a cider environment of sweet and low acidity, meaning plenty of residual fermentable sugars and a pH level above 3.8. It was also believed that *Z. mobilis* was completely immune to sulfur dioxide treatments, even at very high dosages.

It is also often cited that a possible contributing cause is when malolactic fermentation (MLF) takes place, either increasing the bacterial activity or by raising the pH level as malic acid is converted to lactic acid. Recent research has challenged this wisdom, however, showing that *Z. mobilis* can grow in pH as low as 3.2, which is well within most cider pH levels and is at a pH level considered safe.

On the positive side, one study showed that *Z. mobilis* was affected by sulfur dioxide. However, as ethanal increases, its binding effect begins to mitigate the sulfur dioxide effect. It was also found that MLF did not cause cider sickness to develop, though the fact that it lowers pH may be a factor in ciders that are only protected from the disorder by their pH level. Most interesting from this research is a strong correlation between the levels of amino acids (particularly asparagine) in the cider and the development of cider sickness.[30]

Resolution. The research suggests that the amount of residual nitrogen in your cider is the main factor in the development of cider sickness. This would argue for careful use of nutrients. Competition by a more favorable yeast is one possible resolution.

If you have caught it early in the secondary fermentation stage and fermentable sugars are still available, then pitch an aggressive yeast like Lalvin K1-V1116. To help reduce bacterial growth, store the cider during this fermentation at the lowest end of the temperature range for the yeast used, around 50°F/10°C. If it's racking time, bottle and mark these for a minimum of 6 months' storage. Open one up after that and hopefully the smell and flavors will have abated. If not, one last resolution is to pour all the cider back into a carboy, add sugar, and pitch a favorable yeast for a second round of fermentation. You are not shooting for the best cider you have ever made at this point — just drinkable will be a big win. As with all bacterial contaminations, you are going to need to step up your postcleaning regimen after bottling from just sanitizing to sterilizing to kill any bacteria that remain in or on your equipment.

A WHITE FILM (A.K.A. SURFACE YEAST, OR THE FLOWER)

Symptoms. When you peer down into your cider, it looks as if someone has used your cider as their cigarette ashtray, with a shake or two of baby powder to cover up the evidence. Bubbles are often trapped by the film, creating these strange white orbs that look as if they are floating upon a sea of gunk. If you are tempted to scoop it all away, which you will be because finding this thin alien skin on the surface of your cider is scary the first time it happens, you will soon notice that no matter how patient and meticulous you are in your scooping, the following day it's back just as it was. It is a surface yeast. The general term for these yeasts is kahm yeast, and they can be any number of yeasts, such as *Candida mycoderma* or *Mycoderma vini*. They are harmless but don't taste good. The film changes the aroma

to something ranging from musty closet to light vinegar to nail polish remover. If left alone, the alcohol will be broken down, the cider will lose its body, and the smell will remain off permanently.

Cause. Aerobic microbes came in on the fruit, survived due to no sulfur dioxide treatment, and were able to get a toehold because of a wide surface area and sufficient headspace with enough oxygen to thrive. The most common ways this has happened to us are through an airlock that has gone dry or when Christopher has done too heavy a sampling from a carboy, which both increases the surface area of the cider as it is drawn down the neck to the shoulders of the carboy and provides more opportunities to be exposed to oxygen as the cider is repeatedly opened to insert the cider thief. In other words, the causes are completely preventable.

Resolution. There are two basic stages for correcting this issue: remove the film and then change the environment to one that no longer supports its return, which is giving as little possibility for oxygen as possible. A final third method is to treat it.

First the removal. We have used two methods, and a combination of both works best for us. Using a stainless steel spoon, slowly scoop up as much of the film as you can without causing too much to break apart and sink to the bottom. If that happens, switch to the second method, which is to wrap a piece of paper towel around the handle end of a wooden spoon and secure it with a rubber band. Yes, you are making a giant cotton swab–like thing. Gently push the spoon down through the carboy neck and touch the surface of the cider, which will wick the film up into the paper towel. Finish by wiping the neck of the carboy with a paper towel and spray the inside of the neck with a neutral distilled spirit like vodka. Add additional cider to bring up the cider level to as close as you can to the airlock.

Second, change the environment. You can carefully rack into a sanitized carboy. When finished, add additional cider to bring up the cider level to be as close as you can to the airlock. Think of it as going from a halter top to a turtleneck. This often works. If not, the final idea is to get serious.

Unfortunately (if you're a fan of natural cider), it's a good idea at this point (assuming this is a finished cider that was in the storage/maturing stage) to treat it with sulfur dioxide to kill the offending microbes just waiting to recolonize the surface again. Claude Jolicoeur, in his book *The New Cider Maker's Handbook,*[31] suggests applying a dosage of 30 parts per million of sulfite and then spraying another even smaller dose (0.5 percent sulfite to 99.5 percent water) on the neck and small surface of the cider before buttoning it up again.

CONSISTENCY OF EGG WHITES

Symptoms. This infection may not affect the flavor so much, but it is awfully disconcerting to pop the top on a much-awaited cider, and as you pour, out comes something like cider and vegetable oil. It can be worse, tending toward a mixture of cider and egg whites, with the pour including plops of ropy, slimy thickness. Yes, you can drink it, and beyond the obvious off-putting mouth feel there will be a hint of lactic acid, which is the clue leading us to the villain.

Cause. Lactic acid bacteria (LAB) can cause this condition by converting the remaining sugars to lactic acid and polysaccharide gels (think yogurt). We have experienced this during the aging stage or when we have forgotten a cider in its carboy for many months during primary fermentation, leaving it to its lees and increasing its chance of developing these bacteria, which are on the fruit's surface. These LAB are typically killed by the sulfiting process, so you are more likely to see this in wild ciders.

Resolution. You need to break up the gel and kill the bacteria, so it doesn't continue to develop. To do this, transfer the affected cider to an open container and stir vigorously. The cider will begin to lose the feeling of a heavy oil and feel thinner. Next, treat it with sulfur dioxide at a rate of 100 parts per million. Finally, to bring the texture closer to normal, you may want to "fine" with gelatin and bentonite. In this process the gelatin provides a protein with a positive charge to counteract the suspended compounds, which causes them to clump together and drop to the bottom of the carboy. If you add too much gelatin you can create a new type of haze, so the bentonite is also added to counter this, just in case. Once the cider has cleared, we suggest racking into another carboy or glass jug and maturing at least a month to give you time to observe and make sure this infection doesn't come back. If it does, we suggest giving up on the cider; but if it doesn't, move to bottling it.

LICKING A HAMSTER CAGE

Symptoms. Imagine a hamster or mouse cage. Now imagine it hasn't been cleaned in quite a while. Now imagine you are cleaning it, and as part of that process you give the floor of the cage a healthy lick with your tongue. You should be imagining intense unhappiness, which is not far off from the taste of a cider with mousiness. Andrew Lea, in his excellent book *Craft Cider Making,*[32] says that we cannot, under normal conditions, smell this mousiness, we can only taste it, and that what we taste has a lot to do with the pH in our mouth, which varies by individual. People have varying pH levels in their mouths, and we have read that because of that, some people perceive this infection not as rodent droppings but as "freshly baked biscuits," and yes, you just read that correctly. All we have to say about that is that we wish our pH was like that because we have had this infection and it was nowhere near freshly baked biscuits — it was squarely in the ballpark of rodent droppings.

Cause. Either lactobacillus bacteria or *Brettanomyces* yeast are to blame.

Resolution. Unfortunately, there is no good resolution to the mouse. Your only recourse is to dump the cider and sterilize everything that came in contact with the batch. It seems to develop during storage, so proper hygiene during that step, including minimizing headspace, is key to preventing further painful experiences with this one. We have kept bottles of one bad mouse batch for years in the hopes that it would get better but it never did, and we are positive it never would have.

SMELLS AND TASTES LIKE SHERRY

Symptoms. If you have an old bottle of sherry that you have been hanging on to because you might need it for cooking someday, dust it off and pour a small glass. Give it a good sniff and a taste. That's close to what you'll experience with a cider that has oxidized on you, though hopefully the sherry will taste and smell a little stronger. We haven't experienced this in a carboy, but we have with bottles that were forgotten at the back of our shelves or tucked away in a box under the bed (in the days before we had shelves for cider bottles).

Cause. There are a few chemical reactions happening to get to this state. The amount and type of polyphenols in cider vary by variety, so that is a factor. Remember that polyphenols contribute to the color, bitterness, and astringency of ciders — the last two of which we often talk about as the tannins —so here it's the same thing. When these polyphenols are exposed to oxygen, the reaction produces small amounts of hydrogen peroxide, which then oxidizes the ethanol to produce acetaldehydes.[33] Basically, your tannins have oxidized.

Resolution. If you discover it in the bottle and the cider is still drinkable, then your best resolution is to bump all of the bottles from that batch to the front of the line, to be consumed as soon as possible, because the process may not be finished. If it's in the carboy maturing, you could try a second round of fermentation by adding at least 10 tablespoons of sugar to a 3-gallon carboy of cider, then pitch the proofed yeast and go through the process again, making sure to pay close attention to minimizing all contact with air during the process.

SMELLS LIKE NAIL POLISH REMOVER/GLUE

Symptoms. When judging ciders, the aroma of acetone or nail polish remover fall under the "funky, solvent-like" category. It is detectible at a concentration of 15 parts per million in cider. You might hear someone say she detects "EA," which is an acronym for the compound ethyl acetate that is responsible for the aroma. A similar ester, ethyl ethanoate, is fruity at lower concentrations but gluelike in that it reaches concentrations of 200 mg/liter.

Cause. Ethyl acetate is produced at the beginning of the alcoholic fermentation and remains stable through that phase.[34] Different yeasts produce differing levels of ethyl acetate, which is one of the reasons *S. cerevisiae* species are popular among cultured yeast strains — they produce low levels of ethyl acetate.

Resolution. Bring down the concentration of ethyl acetate or ethyl ethanoate to below our threshold for recognizing it, 200 mg/liter. In fact, humans can only perceive ethyl acetate at about 120 mg/liter. By blending the cider with another low-EA cider, you can effectively bring down the amount to either below the drinker's perception or at least to a level that is perceived as pleasant.

SMELLS AND TASTES LIKE VINEGAR

Symptoms. The cider has anywhere from a hint to an intense aroma of vinegar. The taste will also range from just a bit to all it has to produce, depending upon how much of the ethanol has been converted to acetic acid, which is usually in the 5 to 6 percent range for pure apple ciders. Note that this is a different smell and taste from the white film or flower infection also described in this section.

Cause. Vinegar is created by bacteria of the genus *Acetobacter*, which undo all of your yeasts' hard work of producing alcohol from sugar by converting that alcohol to acetic acid. In their defense, this is really where cider goes anyway without our interventions, so it's hard to blame the little guys. Going from fresh cider to hard cider to vinegar is a process, but it's not necessarily a linear one. In fact, the *Acetobacter* bacteria can be converting the ethanol as it's being produced by the various yeasts converting the sugars. These bacteria like a higher temperature, ideally 77°F/25°C to 86°F/30°C, and they need oxygen. Under these conditions, a vinegar mother will eventually form on the top of your cider.

Resolution. In the early stages, you simply have a little acetic acid zip to your cider, and it's something we find very appealing as a change of pace. If it's further along than this, or you aren't into a little sour in your cider, we feel sulfite heroics are unwarranted and it's time to embrace vinegar because you are headed there whether you intended to or not.

APPENDIX 2

Apple Varieties

You can make a good drinking cider from just about any apple (okay, maybe not a Red Delicious), and as evidence you should know that most of the big commercial ciders you enjoy are made from bulk juice, which is a mix of many dessert apples. It is better to balance the sweetness of these apples with some acidity from the culinary apples, and best of all is to add tannic apples to the mix. Often, you use what is easily available to you, so treat these lists as just a guide and nothing more. We make our best cider from unnamed wild apples, which come from trees that likely grew from seeds in past cider pressings that were scattered into the surrounding forest by birds or other critters. In other words, just because an apple variety discovered a world away has made great cider, it doesn't mean you need that variety in order to make great cider.

This list is intended to give you an idea and is just a drop in the bucket of possible varieties.

SINGLE-VARIETY APPLES

These apples have proven they have what it takes to make a great cider without the help of another variety:

- Foxwhelp
- Golden Russet
- Kingston Black
- Pink Lady
- Redstreak
- Ribston Pippin
- Roxbury Russet
- Stoke Red
- That unnamed crab apple you discovered

SWEETS

- Blenheim Red
- Carter's Blue
- Davey
- Envy
- Fuji
- Gala
- Golden Delicious
- Red Delicious
- Smokehouse
- Snow
- Tolman Sweet
- Winter Banana

BITTERSWEETS

- Ashton Brown Jersey
- Brown Snout
- Dabinett
- Medaille d'Or
- Somerset Redstreak
- Tremletts Bitter
- Yarlington Mill

Snow

Roxbury
Russet

Tolman
Sweet

Waltana

Roxbury
Russet

Fearns
Pippin

Smokehouse

Davey

Dutchess of
Oldenburg

Carter
Blue

GLOSSARY

alcohol yield. We admit that a nice hydrometer does the calculations for you, but it's good to understand the math, just in case you break it. Take your desired ABV and multiply that number by 7.5. That gives you the total point gravity drop needed to arrive at that ABV.

For an example that makes the math easy, if we want an 8 percent ABV we would need a 60-point gravity drop ($8 \times 7.5 = 60$), meaning our juice before fermentation would need to be at 1.060 specific gravity or 14.7 Brix. That's fine, but usually we are going the other way, meaning we aren't changing the specific gravity of our juice to reach an exact ABV but we just want to know what the future ABV will be, given the juice we have. The math is nearly the same. Take the starting specific gravity and subtract the predicted ending specific gravity. If you are planning to finish dry to a SG of 1.000, that would be 1.060 minus 1.000, or 60. You then divide your point drop by 7.5 to get the likely final ABV, which in this case would be 8 percent.

amelioration. If your juice has too much sugar for the desired final ABV, you need to dilute the sugars through a process called amelioration, which is done by diluting with water until you reach your desired specific gravity.

Brix. A measurement of the sugar content in juice, measured in degrees Brix from 0 degrees to 30 degrees.

chaptalization. The process of adding additional sugars to juice to raise the specific gravity and thus the potential of higher alcohol levels prior to fermentation.

DAP. This stands for diammonium phosphate and it is an inorganic nitrogen compound that is used as a yeast nutrient. In practice it is often combined with an organic nitrogen compound for a balanced nitrogen feeding. It will make more sense when you understand YAN (opposite page).

fermentable sugars. All of the sugars that occur naturally in apples are fermentable. Pears contain some nonfermentable sugars.

HDPE. This stands for high-density polyethylene and it is used to make everything from small jugs to large but lightweight fermentation tanks.

hydrometer. A glass measurement device used for calculating the specific gravity of a solution like cider.

lactic acid bacteria. Lactic acid bacteria (LAB) are naturally occurring on all fruit and vegetables. Unlike vegetable fermentation, where LAB are the workhorses of the process, in cider they wait on the sidelines until well into the secondary fermentation phase to convert the inherent malic acid to lactic acid. Typical LAB in this process are *Lactobacillus* spp., *Oenococcus oeni,* and *Pediococcus* spp. Unlike many other unwanted bacteria in the cider process that die at lower pH levels, LAB thrive by attaining a pH of 3.5 or below.

malic acid. The predominate acid in apples is malic acid. It is sometimes removed through malolactic fermentation, and sometimes it is added back after fermentation to add acidity. The concentration of malic acid in the apple juice is dependent upon the apple varieties, time of harvest, storage conditions, and climate in the orchard.

malolactic fermentation (MAF). When the naturally occurring malic acid in apple juice is converted by lactic acid bacteria (LAB) into lactic acid and carbon dioxide.

native yeast species. Yeast naturally occurring on apples varies, but scientists generally list the following as common on apples: *Candida, Kloeckera, Metschnikowia, Pichia, Rhodotorula,* and *Torulopsis.*

nonfermentable sugars. Chemical compounds that are not digestible by the natural fermentation process. In the case of apples, there are no natural nonfermentable sugars, so they must be added. Sorbitol, which naturally occurs in pears, is a nonfermentable sugar and is responsible for perry's sweetness when finished to a dry state.

pH. Not many people seem to know exactly what pH stands for and even fewer understand it once they know. We counted ourselves in the not-even-knowing-what-the-acronym-stood-for camp before fermentation bubbled into our lives. It's the negative log of the concentration of hydronium ions in a substance. See, now you know. Specifically, its formula is $pH = -\log[H+]$, but for many of us, that's not all that helpful. Understanding the pH scale is easier and probably more useful for your daily cidermaking activities. The scale ranges from 0 to 14, with 0 being the most acid and 14 being the most alkaline. Neutral is exactly in the middle at 7.

Apple juice is generally in the 3.0 to 4.0 range. In the upper range (above 3.5), there is a higher risk of bacterial growth. This can be adjusted downward with the addition of malic, tartaric, and citric acids.

Saccharomyces cerevisiae. This yeast is so important we are including it in this section all by itself. *Saccharo* means "to come from sugar", while *myces* means "from fungus". Finally, *cerevisiae* means "brewery". Now you see why it is often referred to as brewer's yeast or baker's yeast.

specific gravity (SG). The measure of soluble solids in a liquid that is used to determine a juice's sugar content.

sulfites, SO$_2$, or sulfur dioxide. Sulfur dioxide is a chemical compound that is often referred to by its molecular formula of SO$_2$ or as a sulfite. It smells like a burnt match and acts as a preservative through its antimicrobial properties. It's also an antioxidant.

tannins. These are complex flavonoid polyphenol compounds that occur naturally in the fruit, though at very different levels depending upon the type. They are responsible for astringency and/or bitterness, depending upon their chemical structure and molecular size.[35] If the procyanidin B$_2$ molecule is smaller, it tastes bitter, but if it's larger, it tastes astringent. Tannins are also responsible for the browning of apple flesh as it oxidizes.[36] Tannins are an antioxidant and a preservative.

YAN. Standing for yeast assimilable (or available) nitrogen, this refers to the amount of nitrogen that is available to the yeast during fermentation. Different strains of yeast have different nutritional needs. When those are met, yeast effectively and efficiently transforms juice to cider.

There are two types of nitrogen we are concerned with. Amino acids, which are organic nitrogen compounds, are eaten slowly and steadily by the yeasts. The other type are ammonium compounds, which are inorganic and consumed by the yeasts quickly. Think of it as the difference between a drive-thru fast-food meal you polish off on the way home from work versus a fixed menu of three or five courses that come timed through a delicious hour of eating.

YAN varies by apple variety and even between the same variety grown in different orchards. One thing that you can count on is that the juice from concentrate has the lowest levels of YAN you will come across and therefore may require supplementation.

ENDNOTES

[1] Boyer, Jeanelle Hai, and Rui Hai Liu. "Apple Phytochemicals and Their Health Benefits." *Nutrition Journal* 3, no. 1 (December 2004): 1–15.

[2] *Ibid.*

[3] Hamblin, James. "The Best Probiotics." *The Atlantic*, August 7, 2019. www.theatlantic.com/health/archive/2019/08/probitoic-foods/595687/.

[4] Beech, F. W. "The Yeast Flora of Apple Juices and Ciders." *Journal of Applied Microbiology* 21, no. 2 (1958): 257–66.

[5] Bolarinwa, I., Orfila, C., and Morgan, M. "Amygdalin Content of Seeds, Kernels and Food Products Commercially Available in the UK." *Food Chemistry* 152 (June 2014): 133–139.

[6] Arroyo-López, F. N., et al. "Susceptibility and Resistance to Ethanol in *Saccharomyces* Strains Isolated from Wild and Fermentative Environments." *Yeasts* 27, no. 12 (May 2010): 1005–15. Accessed 8 December 2018.

[7] Zhang, D. and Lovitt, R. W. "Strategies for Enhanced Malolactic Fermentation in Wine and Cider Maturation." *Journal of Chemical Technology and Biotechnology* 81, no. 7 (May 2006): 1130–40.

[8] Baiano, A., Petruzzi, L., Sinigaglia, M., Corbo, M., and Bevilacqua, A. "Fate of Anthocyanins in the Presence of Inactivated Yeasts and Yeast Cell Walls during Simulation of Wine Aging." *Journal of Food Science and Technology* 55, no. 8 (June 2018), 3335–39.

[9] Madera, Roberto R., et al. "Cider Lees: An Interest Resource from the Cider Making Industry." *Waste and Biomass Valorization* 10, no. 6 (December 2017): 1–9.

[10] Madigan, Grace. "Cider Bubbles with Pét-Nat." *CIDERCRAFT*, December 24. 2018. https://cidercraftmag.com/2018/12/24/cider-bubbles-with-pet-nat/.

[11] Herrera, C., de Vega, C., Canto, A., and Pozo, M. "Yeasts in Floral Nectar: a Quantitative Survey." *Annals of Botany* 103, no. 9 (June 2009): 1415–23.

[12] Alonso, S., Laca, A., Rendueles, M., Mayo, B., and Díaz, M. "Cider Apple Native Microbiota Characterization by PCR-DGGE." *Journal of the Institute of Brewing* 121, no.2 (August 2014): 287–89.

[13] Pelliccia, C., Antonielli, L., Corte, L., Bagnetti, A., Fatichenti, F., and Cardinali, G. "Preliminary Prospection of the Yeast Biodiversity on Apple and Pear Surfaces from Northern Italy Orchards." *Annals of Microbiology* 61, no. 4 (December 2011): 965–72.

[14] Valles, B., Bedriñana, R., Tascón, N., Simón, A., and Madrera, R. "Yeast Species Associated with the Spontaneous Fermentation of Cider." *Food Microbiology* 24, no. 1 (February 2007): 25–31.

[15] Cousin, F.J.; Le Guellec, R.; Schlusselhuber, M.; Dalmasso, M.; Laplace, J.-M.; Cretenet, M. "Microorganisms in Fermented Apple Beverages: Current Knowledge and Future Directions." *Microorganisms* 5, no. 3 (July 2017): 39.

[16] Dueñas, M., Irastorza, A., Fernandez, K., Bilbao, A. and Huerta, A. "Microbial Populations and Malolactic Fermentation of Apple Cider using Traditional and Modified Methods." *Journal of Food Science* 59, no. 5 (August 2006): 1060–64.

[17] Cousin, "Microorganisms in Fermented Apple Beverages."

[18] Morrissey, W. F., et al. "The Role of Indigenous Yeasts in Traditional Cider Fermentations." *Journal of Applied Microbiology* 97, no. 3 (2004): 647–55.

[19] Yan, Jia-wei, et al. "The Aroma Volatile Repertoire in Strawberry Fruit: A Review." *Journal of the Science of Food and Agriculture* 98, no. 12 (March 2018): 4395–4402.

[20] Hoelzel, N. "Geobotanical Long-term Monitoring as a Basis for Efficiency Controls in 'Streuobstwiesen' (Scattered Fruit Trees over Extensive Grassland)." *Naturschutz und landschaftsplanung* 31 (January 1999): 147–53.

[21] Rupp, Rebecca. "Applejack. For When Hard Cider Just Isn't Strong Enough." *National Geographic*, October 19, 2015. www.nationalgeographic.com/people-and-culture/food/the-plate/2015/10/19/applejack-for-when-hard-cider-just-isnt-strong-enough/.

[22] Stuckey, Barb. *Taste: Surprising Stories and Science about Why Food Tastes Good.* Atria Paperback, 2012.

[23] *Ibid.*

[24] Williams, A. A. "Flavour Research and the Cider Industry." *Journal of the Institute of Brewing* 80, no. 5 (1974): 455–70.

[25] Guyot, Sylvain, et al. "Multiplicity of Phenolic Oxidation Products in Apple Juices and Ciders, from Synthetic Medium to Commercial Products." *Recent Advances in Polyphenol Research* 1 (July 2008).

[26] de la Roza, C., Laca, A., García, L., and Díaz, M. "Ethanol and Ethyl Acetate Production during the Cider Fermentation from Laboratory to Industrial Scale." *Process Biochemistry* 38, no. 10 (May 2003): 1451–56.

[27] Cousin, "Microorganisms in Fermented Apple Beverages."

[28] Stuckey, *Taste: Surprising Stories and Science.* Atria Paperback, 2012.

[29] Bauduin, R, et al. "Factors Leading to the Expression of Framboise in French Ciders." *LWT — Food Science and Technology* 39, no. 9 (January 2003): 966–71.

[30] *Ibid.*

[31] Jolicoeur, Claude. *The New Cider Maker's Handbook: A Comprehensive Guide for Craft Producers.* Chelsea Green Publishing, 2013.

[32] Lea, Andrew. *Craft Cider Making,* 3rd ed. The Crowood Press, 2017.

[33] Hesseling, Elona. "Aldehydes — From Wine Fault to Fino Sherry." *WineLand*, April 1, 2014. www.wineland.co.za/aldehydes-from-wine-fault-to-fino-sherry/.

[34] de la Roza, "Ethanol and Ethyl Acetate Production."

[35] Lea, *Craft Cider Making.*

[36] *Ibid.*

RESOURCES

BOOKS

Brennan, Andy. *Uncultivated*. Chelsea Green Publishing, 2019.

Carr, Jonathan, Nicole Blum, and Andrea Blum. *Cider House Cookbook*. Storey Publishing, 2018.

Christensen, Emma. *Modern Cider*. Ten Speed Press, 2017.

Cook, Gabe. *Ciderology*. Spruce, 2018.

Crowden, James. *Ciderland*. Birlinn Publishing, 2008.

Jolicoeur, Claude. *The New Cider Maker's Handbook*. Chelsea Green Publishing, 2013.

Lea, Andrew. *Craft Cider Making*. 3rd ed., Crowood Press, 2017.

Otto, Stella. *The Backyard Orchardist*. 2nd ed., OttoGraphics, 1994.

Phillips, Michael. *The Apple Grower*. Chelsea Green Publishing Company, 1998.

Proulx, Annie, and Lew Nichols. *Cider*. 3rd ed., Storey Publishing, 2003.

Ralph, Ann. *Grow a Little Fruit Tree*. Storey Publishing, 2015.

Watson, Ben. *Cider, Hard and Sweet*. 2nd ed., Countryman Press, 2009.

White, April. *Apples to Cider*. Indiana University Press, 2015.

JOURNALS & PODCASTS

Cider Chat

https://ciderchat.com

Cider Chat is probably the best podcast out there for those who love cider. With over 200 episodes archived on their website, you can find any topic, and the host's knowledge and love of cider is infectious

Malus

www.maluszine.com

Malus is a beautiful little quarterly journal that features thoughts, essays, and poetry of cider.

CIDER GROUPS & ASSOCIATIONS

Michigan Cider Association
www.michiganciders.com

Northwest Cider Association
www.nwcider.com

Pennsylvania Cider Guild
https://paciderguild.org

United States Association of Cider Makers
https://ciderassociation.org

Vermont Ice Cider Association
www.vermonticecider.com

NURSERIES

Cummins Nursery Shop
Ithaca, New York
https://shop.cumminsnursery.com

Raintree Nursery
Morton, Washington
https://raintreenursery.com

Stark Bro's
Louisiana, Missouri
www.starkbros.com

TRECO-Oregon Rootstock & Tree Co. Inc.
Woodburn, Oregon
www.treco.nu

Temperate Orchard Conservancy
Mololla, Oregon
www.temperateorchardconservancy.org
Their mission is to save, preserve, and share heirloom and historic varieties of fruits through sales of scion wood. Founded in 2011, three of Nick Botner's friends, Joanie Cooper, Franki Baccellieri, and Shaun Shepherd, and a few others began the Herculean task of grafting every single variety in his orchard.

SUPPLIERS

Bootleg Biology
https://bootlegbiology.com
Bootleg Biology is an open source yeast (and wild bugs) project that has the goal of creating the world's most diverse library of microbes for the creation of fermented foods and beverages. It has wonderful tutorials and the equipment needed to create your own yeast lab.

Cidersupply.com
www.cidersupply.com
They sell a keeving kit as well as pectin methylesterase for keeving under the trademark Klercidre.

GrowlerWerks
Portland, Oregon
www.growlerwerks.com
Among other brewing supplies, GrowlerWerks sells uKeg 1-gallon kegs for carbonation.

Happy Valley Ranch Cider Presses
Paola, Kansas
www.happyvalleyranch.com

Lehman's Hardware
Kidron, Ohio
www.lehmans.com

Midwest Supplies
Minneapolis, Minnesota
www.midwestsupplies.com

Northern Brewer
Roseville, Minnesota
www.northernbrewer.com
True to their name, all three of their stores are in the northern states of Minnesota and Wisconsin.

TRAINING & ADDITIONAL INFORMATION

Cider and Perry Production — A Foundation from Oregon State University
https://pace.oregonstate.edu/catalog/cider-and-perry-production-foundation

Cider Language
https://ciderlanguage.com

Cider Lexicon Project
https://ciderassociation.org/cider-lexicon-project

Certification as a Certified Cider Professional from the United States Association of Cider Makers
https://ciderassociation.org/certification/
You don't have to join the association to take the exam and get the certification, but if you do, there is a discount off the exam fee.

Craft Cidery Startup Workshop from Oregon State University
https://pace.oregonstate.edu/catalog/craft-cidery-startup-workshop

Northwest Cider Association Keeving Blog
www.nwcider.com/keeving

METRIC CONVERSION CHART

TO CONVERT	TO	MULTIPLY
ounces	grams	ounces by 28.35
pounds	grams	pounds by 453.5
pounds	kilograms	pounds by 0.45
teaspoons	milliliters	teaspoons by 4.93
tablespoons	milliliters	tablespoons by 14.79
fluid ounces	milliliters	fluid ounces by 29.57
cups	milliliters	cups by 236.59
cups	liters	cups by 0.24
quarts	milliliters	quarts by 946.36
quarts	liters	quarts by 0.946
gallons	liters	gallons by 3.785
inches	millimeters	inches by 25.4
inches	centimeters	inches by 2.54
inches	meters	inches by 0.0254
feet	meters	feet by 0.3048
feet	kilometers	feet by 0.0003048
miles	meters	miles by 1,609.344
miles	kilometers	miles by 1.609344

BIBLIOGRAPHY

Aleksandra, Štornik, et al. "Comparison of Cultivable Acetic Acid Bacterial Microbiota in Organic and Conventional Apple Cider Vinegar." *Food Technology and Biotechnology* 54, no. 1 (March 2016): 113–19.

Arroyo-López, F. N., et al. "Sutsceptibility and Resistance to Ethanol in *Saccharomyces* Strains Isolated from Wild and Fermentative Environments." *Yeast* 27 (December 2010): 1005–15.

Bauduin, R, et al. "Factors Leading to the Expression of 'Framboise' in French Ciders." *LWT — Food Science and Technology* 39, no. 9 (November 2006): 966–71.

Beckwith, Bob. "Cider Styles: A Primer." *North American Brewers Association*, May 18, 2000. https://north-americanbrewers.org/cider-styles-a-primer/.

Beech, F. W. "The Yeast Flora of Apple Juices and Ciders." *Journal of Applied Bacteriology* 21, no. 2 (December 1958): 257–66.

Bougenies, Nathalie. "The Law on CBD-Infused Alcohol." *Canna Law Blog*, 30 August 2018, www.cannalawblog.com/the-law-on-cbd-infused-alcoholic-beverages/.

CAMRA, "*What is Real Cider?*" *CAMRA*. www.camra.org.uk/cider-articles/-/asset_publisher/7XbXKM0aFgJr/content/what-is-real-cider/.

Cider Consumption Tendencies. "Report On Cider Consumption Globally." *Beverage Trade Network*. https://beveragetradenetwork.com/en/cider-consumption-tendencies-337.htm.

Claus, Michael J., and Kris A. Berglund. "Fruit Brandy Production by Batch Column Distillation with Reflux." *Journal of Food Process Engineering* 28, no. 1 (April 2005).

Crum, Hannah, and Alex LaGory. *The Big Book of Kombucha: Brewing, Flavoring, and Enjoying the Health Benefits of Fermented Tea.* Storey Publishing, 2016.

Delange, N., et al. "Occurrence of Mycotoxins in Fruit Juice and Wine." *Food Control* 14, no. 4 (June 2003): 225–27.

DuPont, Susan M., et al. "Polyphenols from Alcoholic Apple Cider Are Absorbed, Metabolized and Excreted by Humans." *The Journal of Nutrition* 132, no. 2 (February 2002): 172–75.

Explore Yeast, "What is Yeast?" *Explore Yeast*, https://www.exploreyeast.com/article/what-is-yeast/.

Gill, Steven R., et al. "Metagenomic Analysis of the Human Distal Gut Microbiome." *Science* 312, no. 5778 (June 2006): 1355–59.

Grice, Elizabeth A., et al. "Topographical and Temporal Diversity of the Human Skin Microbiome." *Science* 324, no. 5931 (May 2009): 1190–92.

Guyot, Sylvain, et al. "Multiplicity of Phenolic Oxidation Products in Apple Juices and Ciders, from Synthetic Medium to Commercial Products." *Recent Advances in Polyphenol Research* 1 (July 2008).

Hallmann, Caspar A., et al. "More Than 75 Percent Decline over 27 Years in Total Flying Insect Biomass in Protected Areas." *PLOS ONE* 12, no. 10 (October 2017).

Hamblin, James. "The Best Probiotics." *The Atlantic*, August 7, 2019, www.theatlantic.com/health/archive/2019/08/probitoic-foods/595687/.

Hesseling, Elona. "Aldehydes — from Wine Fault to Fino Sherry." *WineLand*, April 1 2014, www.wineland.co.za/aldehydes-from-wine-fault-to-fino-sherry/.

Janik, Erika. *Apple: A Global History.* Reaktion Books, 2011.

Jolicoeur, Claude. *The New Cider Maker's Handbook: A Comprehensive Guide for Craft Producers.* Chelsea Green Publishing, 2013.

Keller, Susanne E., et al. "Influence of Fruit Variety, Harvest Technique, Quality Sorting, and Storage on the Native Microflora of Unpasteurized Apple Cider." *Journal of Food Protection* 67, no. 10 (October 2004): 2240–47.

Lea, Andrew. *Craft Cider Making,* 3rd Edition. Crowood Press, 2017.

Letz, M. L. "Recherches biologiques sur la constitution du Tibi." *Extrait du Bulletin Trimestrel de la Société Mycologique de France* 15, (1899): 68–72.

Madera, Roberto R., et al. "Cider Lees: An Interest Resource from the Cider Making Industry." *Waste and Biomass Valorization* 10, no. 6 (December 2017): 1–9.

Madigan, Grace. "Cider Bubbles with Pét-Nat." *CIDERCRAFT,* December 24, 2018. https://cidercraft-mag.com/2018/12/24/cider-bubbles-with-pet-nat/.

Mangas, Juan, et al. "Changes in the Major Volatile Compounds of Cider Distillates During Maturation." *LWT — Food Science and Technology* 29, no. 4 (1996): 357–64.

Morrissey, W. F., et al. "The Role of Indigenous Yeasts in Traditional Cider Fermentations." *Journal of Applied Microbiology* 97, no. 3 (2004): 647–55.

Otto, Stella. *The Backyard Orchardist: A Complete Guide to Growing Fruit Trees in the Home Garden,* 2nd ed. Ottographics, 1994.

Phillips, Michael. *The Apple Grower: A Guide for the Organic Orchardist.* Chelsea Green Publishing, 1998.

Picinelli, Anna, et al. "Chemical Characterization of Asturian Cider." *Journal of Agricultural and Food Chemistry* 48, no. 9 (September 2000): 3997–4002.

Pidoux, M. "The Microbial Flora of Sugary Kefir Grain (the Gingerbeer Plant): Biosynthesis of the Grain from *Lactobacillus hilgardii* Producing a Polysaccharide Gel." *Journal of Applied Microbiology and Biotechnology* 5, no. 2 (June 1989) 223–38.

Proulx, Annie, and Lew Nichols. *Cider: Making, Using & Enjoying Sweet & Hard Cider,* 3rd ed. Storey Publishing, 2003.

Ridout, Fran, Stuart Gould, Carlo Nunes, and Ian Hindmarch. "The Effects of Carbon Dioxide in Champagne on Psychometric Performance and Blood-Alcohol Concentration." *Alcohol and Alcoholism* 38, no. 4 (July 2003): 381–85.

Rupp, Rebecca. "Applejack. For When Hard Cider Just Isn't Strong Enough." *National Geographic,* October 19, 2015. www.nationalgeographic.com/people-and-culture/food/the-plate/2015/10/19/apple-jack-for-when-hard-cider-just-isnt-strong-enough/.

Stuckey, Barb. *Taste: Surprising Stories and Science about Why Food Tastes Good.* Atria Paperback, 2012.

Symoneaux, R., et al. "Impact of Apple Procyanidins on Ssensory Perception in Model Cider (part 1): Polymerisation Degree and Concentration." *LWT — Food Science and Technology* 57, no. 1 (June 2014): 22–27.

Waldherr, Florian W., et al. "Identification and Characterization of a Glucan-Producing Enzyme from *Lactobacillus hilgardii* TMW 1.828 Involved in Granule Formation of Water Kefir." *Food Microbiology* 27, no. 5 (August 2010): 672–78.

Watson, Ben. *Cider, Hard and Sweet: History, Traditions, and Making Your Own,* 2nd ed. Countryman Press, 2009.

Williams, A. A. "Flavour Research and the Cider Industry." *Journal of the Institute of Brewing* 80, no. 5 (February 1974): 455–70.

Yan, Jia-wei, et al. "The Aroma Volatile Repertoire in Strawberry Fruit: A Review." *Journal of the Science of Food and Agriculture* 98, no. 12 (March 2018): 4395–4402.

INDEX

Page numbers in *italic* indicate photos; numbers in **bold** indicate charts.

C

M

N

DISCOVER THE WORLD OF FERMENTATION
with more books by the Shockeys

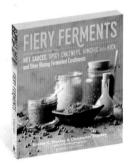

Get to work making your own kimchi, pickles, sauerkraut, and more with this colorful and delicious guide. Beautiful photography illustrates methods to ferment 64 vegetables and herbs, along with dozens of creative recipes.

Expand your fermentation repertoire with more than 70 recipes for spicy sauces, mustards, chutneys, and relishes from around the globe. An additional 40 recipes for breakfast foods, snacks, entrées, and beverages highlight many uses for the hot ferments.

Turn humble beans and grains into umami-rich, probiotic-packed super-foods. Master the fundamentals of fermenting soybeans and rice, then go beyond the customary ingredients with creative alternatives including quinoa, lentils, oats, and more.